INDUSTRIAL APPLICATIONS OF SURFACTANTS

Special Publication No. 59

Industrial Applications of Surfactants

The Proceedings of a Symposium organised by the North West Region of the Industrial Division of the Royal Society of Chemistry

University of Salford, 15th—17th April 1986

Edited by
D. R. Karsa
Lankro Chemicals Limited, Manchester

ROYAL
SOCIETY OF
CHEMISTRY

British Library Cataloguing in Publication Data

Industrial applications of surfactants : the proceeding of a symposium organized by the North West Region of the Industrial Division of the Royal Society of Chemistry, University of Salford, 15–17th April 1986.—(Special publications; no. 59)
1. Surface active agents
I. Karsa, D. R. II. Royal Society of Chemistry. *Industrial Division. N.W. Region*
III. Series
668′.1 TP994

ISBN 0-85186-666-2

Reprinted 1988

Published by The Royal Society of Chemistry,
Burlington House, London, W1V 0BN

Printed in Great Britain by
Whitstable Litho Printers Ltd, Whitstable, Kent

Introduction

Throughout the 1970s and early 1980s, there has been a growing awareness of the diversity of applications for surface active agents outside the area of domestic detergents and personal care products. This 'industrial' area consumes approximately 40% of all surfactants produced, many of them being sold as part of formulated 'effect' or 'performance' chemicals. In many cases the end-user purchases a 'dewatering aid', a 'defoaming agent' or a 'dispersant', frequently not fully realising that the product is based on surfactant components.

It is difficult to find an industry which does not directly or indirectly consume surfactants, yet many international conferences and meetings have tended to concentrate on raw materials and surfactant components for use in primarily domestic detergents and household products.

The North West Region of the Industrial Division of the Royal Society of Chemistry considered that the time was appropriate to arrange an international conference which concentrated on the 'Industrial Applications of Surfactants', in order to foster an awareness of the many uses of these materials. A two-day conference can only cover a limited number of application areas; however, it is hoped that the conference, which was so well supported by over 350 delegates, and this monograph will provide useful and stimulating background reading to producers and formulators of surfactants alike.

D. R. KARSA

Acknowledgements

The North West Region of the Industrial Division of the Royal Society of Chemistry wishes to express its gratitude to the following co-sponsors for their generous support:

Akzo Chemie BV

I.C.I. p.l.c., Petrochemicals & Plastics Division

Lankro Chemicals Limited

Shell Chemicals (UK) Limited

Contents

Industrial Applications of Surfactants - an Overview — 1
By D. R. Karsa

Specific Properties of Silicone Surfactants and Their Applications — 24
By G. L. F. Schmidt

Surfactants Widely Used in the Concrete Industry — 33
By B. El-Jazairi

Antifoams in Gas-Oil Separation — 48
By I. C. Callaghan, S. A. Hickman, F. T. Lawrence, and P. M. Melton

Anionic Surfactants — 58
By N. E. Jones and J. W. S. Patterson

Surfactants and Emulsion Polymerisation — 73
By R. Groves

Interaction of Fire Fighting Foams with Burning Hydrocarbons — 90
By A. A. Briggs

Applications of Surfactants and Polymers in Pesticidal Suspension Concentrates — 102
By Th. F. Tadros

Surfactants in Enhanced Oil Recovery — 126
By T. F. Ling, H. K. Lee, and D. O. Shah

Industrial Applications of Nonionic Surfactants — 179
By M. Hellsten

Surfactants in Textile Processing — 195
By K. R. F. Cockett

Formulation of Industrial Hygiene Products: Art or Science — 208
By J. R. Sutton

Coal-Water Slurries — 235
By G. Bognolo

Cationic Surfactants — 250
by A. D. James, P. H. Ogden, and J. M. Wates

Cationic Surfactants in Mineral Processing — 269
By H. L. Shergold

Organophilic Clays 289
By J. P. Tatum

Antistatic/Conductivity Additives in Paint and Plastics 307
By. B. Davis

Industrial Applications of Co-polymer Surfactants 318
By M. A. Plant

Fluorochemical Surfactants 333
By M. C. Allison

Subject Index 343

Industrial Applications of Surfactants - an Overview

By D. R. Karsa
LANKRO CHEMICALS LIMITED, PO BOX 1, ECCLES, MANCHESTER M30 OBH, UK

Surface Active Agents, or 'Surfactants', impact on all aspects of our daily life, either directly in household detergents and personal care products or indirectly in the production and processing of the materials which surround us. The domestic product area is by far the largest consumer of surfactants. However, as will be amply illustrated by this conference, there are large volumes of surfactants utilised in numerous non-domestic applications.

Of the 4 million tonnes of surfactants used in the United States, Japan and Western Europe, almost half are used in industrial applications.

With few exceptions, in the 'industrial applications' the surfactants are used either as essential additives or processing aids and, in many cases, as only a minor part of a particular system or formulation. Here the surfactant may be regarded as a true 'performance' or "effect" chemical where detailed knowledge of a surfactant's composition and surface active properties is essential for selection of the optimum product for a particular end-use. In many cases a single surfactant cannot provide all the properties required and two or more must be formulated to afford the desired effect. In other areas where conflicting properties are sometime unavoidable the surfactant selected is a compromise where, on balance, the properties achieved are the most beneficial in the circumstances.

The discussion of such a diverse topic as surfactant applications must of necessity be a general overview. Therefore, it has to be assumed that all readers have a basic understanding of the nature of surfactants and the fundamental properties they exhibit. In brief, we are considering the behaviour of amphiphilic molecules consisting of a hydrophilic (water-compatible) head group and a hydrophobic (water-repellant) hydrocarbon tail (Figure 1).

FIGURE 1

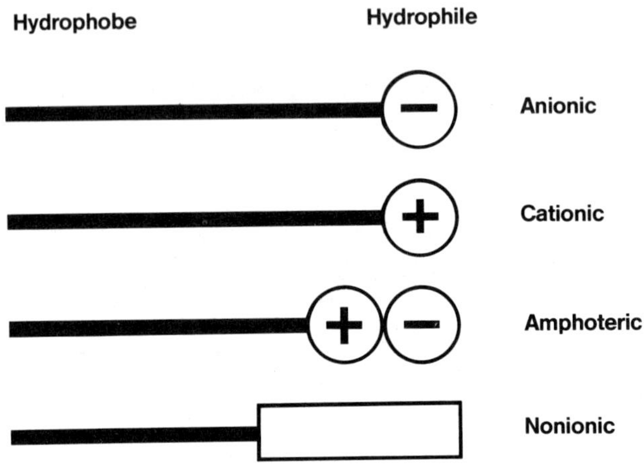

This incompatibility within the same molecule is the cause of the unique behaviour of the surfactant molecule, the molecule orientating in such a way as to place its hydrophobic group in a hydrophobic environment. Hence, we see the formation of an absorbed layer at an interface and the agglomeration of surfactant molecules into spherical,cylindrical or lamellar micelles above the CMC (critical micelle concentration).

The interfacial and solution behaviour of surfactants leads to the following key surface-active properties.

Surfactant properties

Surfactant properties
Emulsification/Demulsification
Wetting/Re-Wetting
Foaming
Dispersing
Defoaming
Detergency
Solubilising

In many applications, a combination of several of these properties is required and various examples will be considered later.

Surfactant types

Examination of the anionic, nonionic, cationic and amphoteric ranges of surfactants shows them to consist of the following major species:

Anionic surfactants

Alkyl Benzene Sulphonates	Alpha-sulphonated Fatty Acids (and Esters)
Alcohol Sulphates	Sulphonated Fatty Acids & Esters
Alcohol Ether Sulphates	Mono - and Di-Ester Sulphosuccinates
Alkyl Phenol Ether Sulphates	Sulphosuccinamates
Fatty Amide Ether Sulphates	Petroleum Sulphonates
Alpha-olefin Sulphates	Phosphate Esters
Paraffin (Alkane) Sulphonates	Ligno Sulphonates

Of the Anionic surfactants, alkylbenzene sulphonates continue to be the major product. Fatty alcohol sulphates and fatty alcohol ether sulphates have gained an increasing market share and, in the domestic product area in particular, alkane (or paraffin) sulphonates and α-olefin sulphonates have made significant penetration of the market.

Nonionic surfactants

Alkyl Phenol Ethoxylates
Fatty Alcohol Ethoxylates
Ethylene Oxide - Propylene Oxide co-polymers
Fatty Amine Ethoxylates
Fatty Acid Ethoxylates (Peg Esters)
Fatty Acid EO/PO Co-polymers
Alkanolamides & Ethoxylates
Sorbitan Esters & Ethoxylates

Cationic surfactants

Quaternary Ammonium Salts
Imidazolinium Salts

Amphoteric surfactants

Betaines
Imidazolines

The Nonionic market continues to be dominated by ethylene oxide adducts of alkylphenols and fatty alcohols. Production of detergent-chain-length fatty alcohols from both natural and petrochemical bases has now exceeded 750,000 tonnes in the West with usage of alkylphenol ethoxylates in some application areas being environmentally less acceptable because of their slower rate of biodegradation and concern regarding the toxicity of phenolic residues.

The Cationic surfactant market is still dominated by quaternary ammonium salts of the di-stearyl di-methyl ammonium chloride type which are used in domestic fabric conditioners, although di-tallow-based imidazolinium salts have made some inroads into this market. Shorter-chain quaternaries are used for their bactericidal effect.

Of the approximate 4 million tonnes of surfactants consumed in the USA, Japan and Western Europe in 1982, well over two-thirds of the volume comprised five main surfactant groups - alkylbenzene sulphonates, lignin sulphonates, fatty alcohol ethoxylates, fatty alcohol ether sulphates and alkyl phenol ethoxylates (Table I).

TABLE I PRODUCTION USA/JAPAN/WESTERN EUROPE(te/a) 1982

SURFACTANT	K TONNES	%
ALKYL BENZENE SULPHONATES	1,100	28.0
LIGNIN SULPHONATES	600	15.3
FATTY ALCOHOL ETHOXYLATES	500	12.8
FATTY ALCOHOL ETHER SULPHATES	360	9.2
CATIONIC SURFACTANTS	310	7.9
ALKYLPHENOL ETHOXYLATES	310	7.9
FATTY ACID ESTERS	280	7.1
FATTY ALCOHOL SULPHATES	220	5.6
ALKANE AND ALPHA-OLEFIN SULPHONATES ETC.	140	3.6
FATTY ACID AMIDES AND ALKANOLAMIDES	100	2.6
TOTAL	3,920	100.0

ANALYSIS OF THE SURFACTANT MARKET (AREA/VOLUME/INDUSTRY)

Statistics on the detailed consumption and application of surfactants are not readily available on a worldwide basis and, at best, are available from the United States, Western Europe and Japan. Data is accumulated by international surfactant organisations such as C.E.S.I.O. in Europe, by major corporations with specific sector interests or by Market Research companies. Even so, all figures quoted should still be viewed as only an approximate guide to consumption. In a product group so diverse as surfactants, where many products are sold as aqueous solutions or in-part used as intermediates to produce other surfactants (eg. fatty alcohol ether sulphates produced from fatty alcohol ethoxylates), double counting cannot be avoided. Likewise, most market surveys are targetted at a specified selection of the major application areas with some remaining segments not being considered.

Worldwide annual consumption of synthetic surfactants and soaps is approximately 13 million tonnes of which almost 40% comprises synthetic surfactants - about 5 million tonnes.

Professor Werdelmann's opening paper at 1984 World Surfactants Congress (Munich) provides us with an excellent analysis of surfactant consumption in the 'West' in 1982 (see Table II).

TABLE II

1982 Surfactant consumption (USA, Western Europe & Japan)

	K Tonnes	%
Detergents + Cleaners (includes some soap)	1,950	49.4
Cosmetics + Pharmaceuticals	310	7.9
Textiles + Fibres	700	17.7
Mining, Flotation, Petroleum Production (Includes some Soap)	300	7.6
Paints, Lacquers, Plastics (includes some soap)	210	5.3
Food Industry (Includes both Additives + Cleaning products)	200	5.1
Pulp + Paper	110	2.8
Agrochemical	105	2.7
Leather + Fur	60	1.5
Total	3,945	

This is also expressed in percentage terms in the following 'block' diagram (Figure 2).

FIGURE 2

1982 Surfactant consumption (USA, Western Europe & Japan)

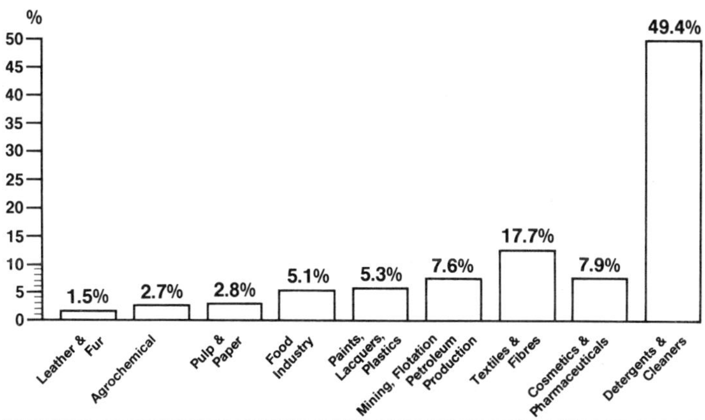

Comparative percentage by volume usage in 1982 is given for the United States, Western Europe and Japan in Figure 3 below:

FIGURE 3

Regional surfactant production 1982

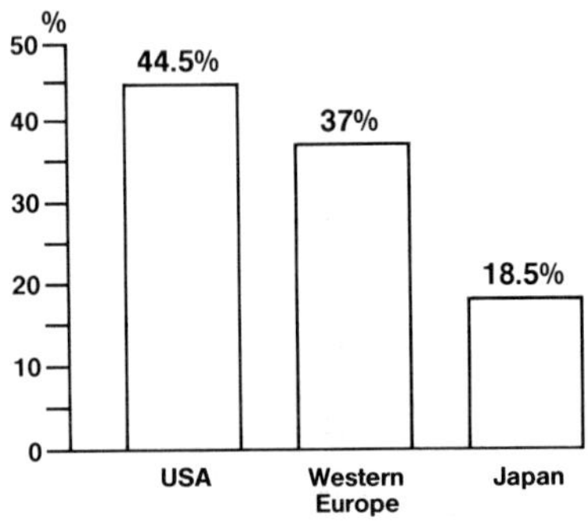

More recent surveys (2) (3) have focussed on the United States and Western Europe. Unfortunately, the market segments surveys are not exactly the same and the USA figures in brackets in Table III are more directly comparable with the European figures opposite. The USA surfactant consumption is higher than Western Europe largely due to differences in technology, especially in the food, paper, plastics and elastomers area and in oil-field chemicals.

TABLE III

1985 Surfactant consumption (excluding soap) – in K Tonnes

	Western Europe	U.S.A.	
Household Products	850	710	
Personal Care Products	96	91	
Industrial & Institutional Cleaning	137	164	
Miscellaneous Industrial Uses (Majors: Textiles/Paint/ Plastics & Elastomers/Cement)	506	1,318	(789)
Oilfield	11	181	
Total	1,600	2,464	(1929)

This is also illustrated in the 'block' diagram below (Figure 4).

FIGURE 4

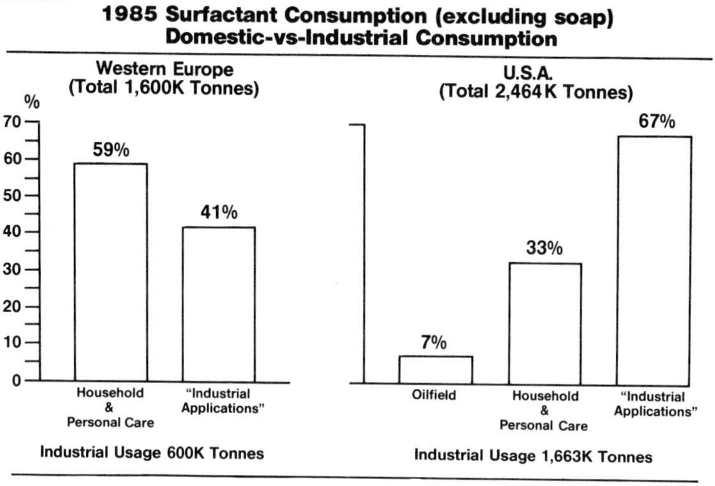

Examination of Western European consumption in 1985 by surfactant type affords the following data (Figure 5).

FIGURE 5

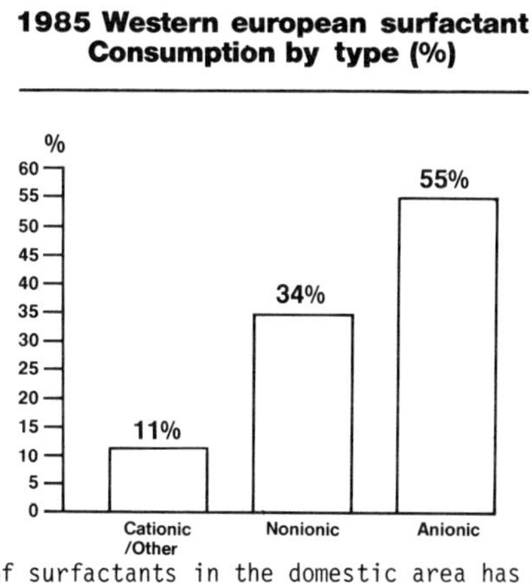

The use of surfactants in the domestic area has shown little variation in consumption over recent years. 'Industrial' usage can show greater variance depending on the fortunes of a particular industry or segment within an industry. Examples illustrating this include the demand for oil-field chemicals depending on the supply and demand for oil, the seasonal demand for surfactants for agrochemical use, which also depends on the prevailing natural conditions, and the fortunes of the textile and fibre areas, which are generally depressed but still with some growth areas. This can be illustrated by the UK 1985/1986 sectoral production forecast (Figure 6).

FIGURE 6

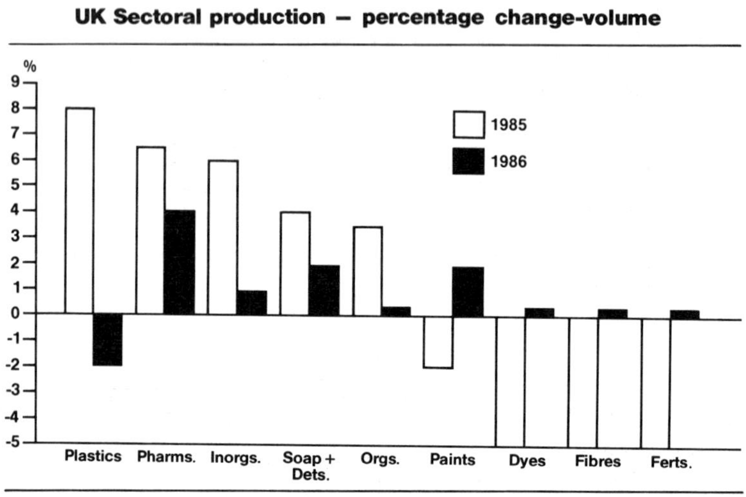

INDUSTRIAL APPLICATIONS

For simplicity, the application areas have been broken down into the following twelve industry sectors. Examples of applications and products are tabulated for each sector.

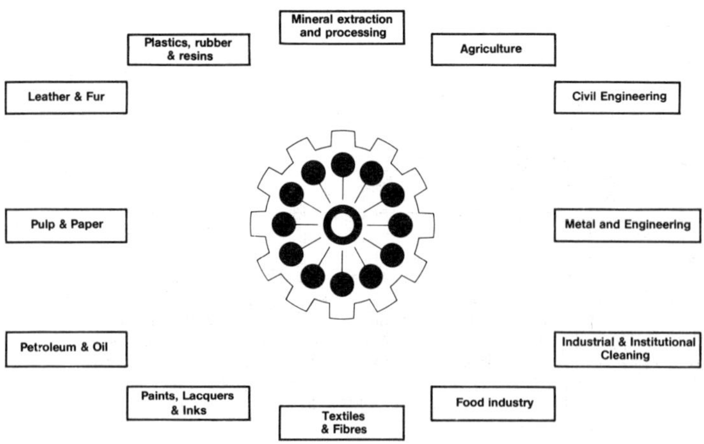

The end-uses are not comprehensive but selected to illustrate the diversity of use.

Industrial and institutional cleaning

Low foam wetters for dishwashing powders and rinse aids	Phosphate esters
	Ethylene oxide - propylene oxide co-polymers
	Fatty alcohol alkoxylates
Hard surface cleaners	Alkanolamides
	Alkyl sulphates
	Fatty alcohol ethoxylates
	Sulphonated fatty acid salts
	Phosphate ester hydrotropes
	Complex amphoterics
Sanitisers	Fatty alcohol ethoxylates
	Quaternary ammonium salts
Laundry detergents	Alkyl benzene sulphonates
	Alcohol ethoxylates
Bottle washing	Fatty alcohol alkoxylates
	Ethylene oxide- propylene oxide co-polymers
	Phosphate ester hydrotropes
Dry cleaning	Sodium dialkyl sulphosuccinates
	Phosphate esters
	Amine salts of alkyl benzene sulphonates
Iodophors/dairy and brewery cleaning	Alkyl phenol ethoxylates
	Phosphate esters
	Fatty alcohol ethoxylates
	Ethylene oxide- propylene oxide co-polymers
	Complex amphoterics
Hand cleansing gels	Alkanolamides
	Alkyl benzene sulphonates amine salts
	Fatty alcohol ethoxylates
	Nonyl phenol ethoxylates
Carpet shampoos	Alkyl sulphates
	Alkanolamides
	Sulphosuccinates
Wax and silicone polishes	Fatty alcohol ethoxylates
	Amine ethoxylates

Examples include:

Cleaning-in-place - cleaning and sanitising of fixed pipework and equipment in breweries, dairies, etc., where low foam, excellent wetting and detergency are a prerequisite.

Machine dishwashing - Rinse aids - nonionic additives to ensure glassware and crockery are left clean and streak-free after washing. Again excellent wetting and very low-foam characteristics are essential over a wide temperature range.

Here the need for a biodegradable product is a further essential requirement and it is unavoidable to have to compromise between this and the low-foam/good wetting characteristics also required.

Textiles and fibres

Detergents, scouring and wetting aids	Nonyl phenol ethoxylates and solutions
	Fatty alcohol ethoxylates
	Ethylene oxide - propylene oxide co-polymers
	Alkyl benzene sulphonates and detergent concentrates
Antistatic agents	Polyglycol esters
	Alkanolamides
	Amine ethoxylates
	Phosphate esters
	Quaternary ammonium salts
Dyestuff auxiliaries	
1 Azoic printing aids, sulphur dyestuff additives	Alkyl benzene sulphonates
2 Dye levelling, dispersing	Solutions of nonyl phenol ethoxylates
	Phosphate esters
3 Pigment dispersants	Polyglycol esters
Fabric softeners	Quaternary ammonium salts
Fibre lubricants	Polyglycol esters
	Ethylene oxide - propylene oxide co-polymers
Kier boiling assistants	Alkyl benzene sulphonates
	Phosphate esters
	Fatty alcohol ethoxylates
	Alkanolamides
	Fatty alcohol ethoxylates
Printing aids	Alkanolamides
Self-emulsifiable oils	Fatty alcohol ethoxylates
	Polyglycol esters
	Nonyl phenol ethoxylates
Shrinkproofing agents for wool	Organo-thio sulphates

Examples include:

Levelling agents in dyeing

- used to ensure evenness of dyeing.

Fibre lubricants

- additives selected for their fibre-to-fibre and fibre-to-metal frictional and cohesive properties to ensure effective fibre production and processing.

Shrink-proofing of wool

- an example of a highly sophisticated and reactive anionic surfactant system capable of both physical and chemical bonding to wool fibres.

Agriculture

Emulsification of pesticide and herbicides in solvent	Calcium and amine salts of alkyl benzene sulphonates
	Nonyl phenol ethoxylates
	Phosphate esters
	Castor oil ethoxylates
	Ethylene oxide-propylene oxide co-polymers
Dispersants for suspension concentrates	Phosphate esters EO/PO co-polymers Fatty acid sulphonates Ligno sulphonates
Wettable powders	Ligno sulphonates Naphthalene sulphonic acid-formaldehyde condensates Alkyl naphthalene sulphonates
Water-dispersible granules	Naphthalene sulphonic acid - formaldehyde condensates Alkyl phenol - formaldehyde resin ethoxylates Alkyl benzene sulphonate powders
Wetting and spreading	Nonyl phenol ethoxylates
	Sulphonated esters
	Alkyl benzene sulphonates (for powders)
	Fatty alcohol ethoxylates
Oil emulsions	Fatty alcohol ethoxylates
	Polyglycol esters
Orchard sprays	Fatty alcohol ethoxylates
	Ethylene oxide - propylene oxide co-polymers
Sheep-dip emulsifiers	Alkyl benzene sulphonates
	Alkyl phenol ether sulphonates
	Phosphate esters
Cattle-bloat remedies	EO/PO co-polymers

Examples include:

Emulsifiable concentrate (EC)

- Solvent/surfactant/toxicant concentrate exhibiting excellent flash emulsification and emulsion stability.

Application of EC by land and aerial spraying.

Wetter-spreaders

- wetting agents to optimise activity of a herbicide.

Civil engineering

Bitumen additives to give adhesion to wet aggregate	Amine ethoxylates
	Quaternary ammonium salts
	Imidazolines
Concrete and mortar plasticisers	Nonyl phenol ethoxylates
	Alkyl benzene sulphonates
	Naphthalene sulphonic acid - formaldehyde condensates
Foaming of concrete or gypsum	Alkyl benzene sulphonates
	Fatty alcohol ether sulphates
	Alkyl phenol ether sulphates
	Betaines
Foaming of urea-formaldehyde resins for cavity-wall insulation	Alkyl benzene sulphonic acids and salts

Example: Concrete superplasticiser

- the use of products such as naphthalene sulphonic acid - formaldehyde condensate derivatives to modify the flow properties, workability and setting time of concrete.

Metal and engineering

Wetting in pickling and plating baths	Phosphate esters
	Fatty acid sulphonates
	Alkanolamides
	Nonyl phenol ethoxylates
Cutting and drilling oils	Alkanolamides
	Phosphate esters
Emulsifiers for solvent degreasers	Fatty alcohol ethoxylates
	Nonyl phenol ethoxylates
	Ethylene oxide-propylene oxide co-polymers
	Amine ethoxylates
	Polyglycol ethers
Dewatering fluids	Amine ethoxylates
	Alkanolamides
Hydraulic fluids	Polyglycols
	Ethylene oxide-propylene oxide co-polymers

Example: Cutting oil emulsifiers

Pulp & paper

Low foam wetting in paper pulping	Polyglycol esters
De-ink of waste	Nonyl phenol ethoxylates
	Fatty alcohol ethoxylates
Defoaming of pulp	Polypropylene glycols
	Polyglycol esters
Wax emulsifiers for chipboard manufacture	Fatty alcohol ethoxylates
	Amine ethoxylates
Viscose production	Alkanolamides
	Phenol ethoxylates
Fluff pulp bulking aids	Quaternary ammonium salts
	Alkyl phenol and fatty alcohol alkoxylates
Dispersants for inorganic coating fillers	EO/PO co-polymers
	Alkyl phenol ether sulphates
	Phosphate esters

Example:

Wash de-inking - the use of nonionic surfactants in the de-inking of newspaper and magazine waste to enable recycling for tissue and board production.

Food

Defoamers for the production of yeast and the extraction of sugar from beet	Polyglycol-based low toxicity antifoams
	Polyglycol esters
	EO/PO co-polymers
Wetters for fruit processing	Alkyl sulphates
	Ethylene oxide-propylene oxide co-polymers
Food grade emulsifiers	Glycerol mono-, di- and tri-esters
	Sorbitan esters and ethoxylates
	Lecithin
Cleaning & sanitising of brewery/dairy equipment etc.	Alkyl phenol ethoxylates
	Phosphate esters
	EO/PO co-polymers
	Quaternary ammonium salts

Examples include:

Food additives - food grade emulsifiers

Extraction of sugar from beet - nonionic surfactants for the production of defoaming agents.

Mineral extraction and processing

Coal-water mixtures	Nonyl phenol ethoxylates
	Naphthalene sulphonic acid - formaldehyde condensates
Coal-oil mixtures	Alkyl benzene sulphonates
	Sulphosuccinates & sulphonated fatty acid esters
	Quaternary ammonium salts
	Amine ethoxylates
Dewatering of mineral slurries	Sulphosuccinates
	Phosphate esters
Flotation	Alkyl aryl sulphonates
	Sulphosuccinamates
Freeze conditioning agents /Side-car release agents	Various alkylene oxide derivatives
Anti-dusting agents	EO/PO co-polymers

Examples:

Chemicals for coal

Mineral pigments and fillers - use of dispersing agents for the production of high-solids/low-viscosity slurries to aid transportation and formulation.

Petroleum and oil

Oil spill chemicals	Polyglycol esters
	Sulphosuccinates
	Sorbitan esters and ethoxylates
Enhanced oil recovery (EOR)	Alkyl and alkyl phenol ether carboxylates and ether sulphonates
	Petroleum sulphonates
	Alkyl xylene sulphonates
Demulsifiers /Desalters	Alkyl phenol-formaldehyde resin ethoxylates
	EO/PO co-polymers
	Alkyl benzene sulphonates
Drilling mud chemicals	Naphthalene sulphonic acid -formaldehyde condensates
	Nonyl phenol ethoxylates
	Fatty alcohol and fatty alcohol ether sulphates
	Imadazolines
	Calcium dodecyl benzene sulphanate
	Alkyl phenol ether sulphates
	EO/PO co-polymers
	PEG esters

Examples include:

Oil-spill dispersants - use of surfactants with low fish toxicity to disperse oil at sea or on beaches.

Oil-field chemicals - demulsifiers, desalters, corrosion inhibitors, EOR chemicals, oil-well cementation,etc.

Paints, lacquers and inks

Pigment dispersion and wetting	Sodium naphthalene sulphonate formaldehyde condensates
	Alkyl sulphates
	Sodium dialkyl sulphosuccinates
	Nonyl and polysubstituted phenol ethoxylates
	Phosphate esters
Flow promoters	Amine ethoxylates
Viscosity modifiers	Fatty alcohol ethoxylates
Resin emulsifiers	Fatty alcohol ethoxylates

Examples include:

Paint production

- use of surfactants as pigment wetters and dispersants, as emulsifiers in the production of the base latex and as defoaming agents to control foam during paint production and application.

Leather and fur

Wetting/penetration	Nonyl phenol ethoxylates
	Fatty alcohol ethoxylates
	Sulphated monoesters of fatty acids
	Sulphated oils
	Fatty acid esters of long-chain fatty alcohols
Soaking and degreasing	Nonyl phenol ethoxylates
	Alkyl sulphates
	Alkyl benzene sulphonates
Pigment pastes	Fatty alcohol ethoxylates
	Nonyl phenol ethoxylates

Plastics, rubber and resins

Primary emulsifiers for the production of emulsion polymers	Branched- and straight-chain alkyl benzene sulphonates
	Fatty alcohol ethoxylates
	Nonyl and alkyl phenol ethoxylates
	Solutions of alcohol and nonyl phenol ethoxylates
	Alkyl sulphates
	Alkyl ether sulphates
	Alkyl phenol ether sulphates
	Phosphate esters
	Ethylene oxide - propylene oxide co-polymers
	Fatty acid sulphonates
	Sulphosuccinic acid and sulphosuccinamic acid derivatives
Tufted carpet industry **1 Foaming agents for secondary foam backings**	Sulphosuccinamic acid derivatives
	Alkyl sulphates
	Complex amphoteric surfactant
2 Wetting agents for primary backing compounds	Nonyl phenol ethoxylates
	Sodium dialkyl sulphosuccinates
	Nonyl phenol ether sulphates
3 Dispersing agents for cure pastes	Sodium naphthalene sulphonate -formaldehyde condensates
4 Odour control agents	Amine ethoxylates and oxides
The coating of textiles, leather and leather cloth and binding of non-wovens	Sulphosuccinamic acid derivatives
	Alkyl sulphates
Modification of plastisol viscosity	Fatty alcohol ethoxylates
	Sodium dialkyl sulphosuccinates
Antistatic agents for PVC	Alkanolamides
	Polyglycol esters
	Quaternary ammonium salts
	Alkyl benzene sulphonates
	Ethylene oxide - propylene oxide co-polymers
	Nonyl phenol ethoxylates

Examples include:

Antistatic agents for plastics

Emulsifiers for emulsion polymers

Latex carpet backing - foaming agents, dispersants, wetting agents.

The preceding lists of applications are by no means comprehensive but illustrate the many and varied applications of surface-active agents.

Selection of products for a specific end-use depends in some cases on toxicological and environmental considerations as well as upon composition - performance parameters. Thus, as we have seen, a low-foam nonionic surfactant for an industrial rinse aid must also be biodegradable according to EEC legislation. Likewise a surfactant used in agrochemical toxicant formulations should ideally comply with the United States E.P.A. Regulations, an oil-spill chemical should pass official fish toxicity and performance tests and surfactants to be used as additives in plastics, paper and board in indirect or direct food contact will be limited by chemical type and addition level according to various regulations.

Where surfactants are used as processing aids, selection frequently depends on an intimate knowledge of the structure and by-products in a particular surfactant. All surfactants, at best, are a complex mixture of surface-active components and other minor species. Thus a simple nonionic ethoxylate is always a Gaussian distribution of ethoxylates with an average number of ethylene oxide units by which the product is defined. It will contain free polyethylene glycol, minor amounts of water and some free base hydrophobe if it is a lower ethoxylate. Similarly a sulphated or sulphonated product is always, at its simplest, a mixture of varying chain-length sulphates/sulphonates with varying levels of unsulphated matter and inorganic sulphate. In some applications, the levels of these by-products can prove to be a critical performance parameter. It is not possible to illustrate all these features in the time remaining, but the following example has been selected to demonstrate how structure and performance can be critically related.

Example : Phosphate ester surfactants for use as
(a) Dispersants for inorganic pigments
(b) Hydrotropes.

Phosphate ester surfactants are a classic example where the end-user may have great difficulty understanding performance differences between competitive products of nominally the same chemical type. Likewise the surfactant producer needs to appreciate the subtleties of phosphation techniques, as totally different products result from different reactant ratios and the type of phosphating agent employed.

If we first consider the methods of phosphation this will assist us to understand the complex structures of the resultant phosphate esters.

The organic phosphates under discussion are those normally prepared by reacting an alcohol or an ethoxylate with the so-called tetraphosphoric acid (TPA) or with phosphorus pentoxide. The chemistry of the formation of these phosphates is complicated making it difficult to assign precise formulae to their composition. It is certain, however, that the composition and properties of the alkyl acid phosphates are significantly different when prepared using different phosphating agents.

Tetraphosphoric acid (TPA) route

Tetraphosphoric acid is essentially a mixture of polyphosphoric acids of the empirical formula:

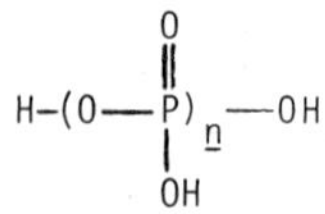

and said to possess the following distribution:

n	1	2	3	4	5	6	7	>7
%	5	17	16.5	14.2	12.0	8.7	2.2	18.9

The reaction between a hydroxylic compound and TPA proceeds via the attack of the -OH on the —P—O—P— bond:

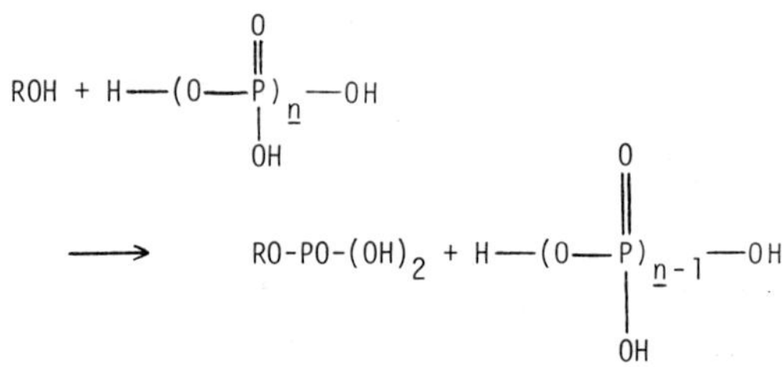

However, if during the course of the reaction a 'diprotic species' (*n* = 2) was formed (via attack from both ends of the chain) then this molecule would not cleave further. The resulting product therefore contained free ortho phosphoric acid, monoalkyl ester and dialkyl pyrophosphate ('diprotic species'). The evidence for this postulate was supported by the absence of di-ester and middle-group phosphorus in analysis by NMR.

Analysis of products prepared using TPA as phosphating agent tends to confirm the proposed mechanism. Calculations using the results from titrations using standard alkalis and based on the above premises give the following approximate results for a typical phosphated alcohol:

		%
Monoalkyl phosphates:	$O = P(OH)_2 OR$	68
Dialkyl pyrophosphate:	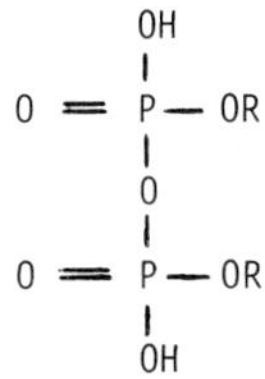	2
Ortho phosphoric acid:	$O = P(OH_3)$	10
Free alcohol:	ROH	20

As can be seen, there is a preponderance of mono-ester, a significant amount of free ortho phosphoric acid and only a small amount of the 'diprotic' pyrophosphate.

Phosphorus Pentoxide Route

The alkyl acid phosphates prepared using phosphorus pentoxide as phosphorylating agent are less completely defined. The structure of phosphorus pentoxide is still open to some discussion. There is good evidence to show that some forms of the oxide exist as tetrahedral structures of P_4O_{10} but the normally commercially available material probably also contains high and very high polymeric material. It is therefore advisable, and convenient, to retain the formulation P_2O_5 for general usage.

It can be assumed that the alcoholysis of highly polymerised chain polyphosphoric oxides will generally follow the paths outlined for TPA, but the alcoholysis of the tetrahedral P_4O_{10} and other ring polyphosphoric oxides can theoretically follow a number of different reaction paths giving rise to a number of different product mixes. This is reflected in the fact that if the guidelines used for alcoholysis of TPA are used here it is difficult to reconcile totally the theoretical compounds with the results obtained by conventional analysis. Typical results can be summarised as:

	%
Monoalkyl phosphate: $O=P(OH)_2$ OR	35
Ortho phosphoric acid: $O=P(OH)_3$	1
Free alcohol : ROH	22
'Polyphosphates' by difference	42

Whatever the true interpretation of the results they indicate that the composition of a phosphate prepared from phosphorus pentoxide differs significantly from that prepared from TPA. When the results are compared with those obtained from a TPA-based product, it will be seen that whilst the free alcohol content is essentially the same the monoester content has been considerably reduced at the expense of an increase in the 'diprotic' or polyphosphate content. These differences are usually carried through into the practical uses of the products where the different compositions are found to be more or less effective in different applications.

(a) Phosphate esters as dispersants for inorganic pigments

In a particular pigment-dispersing application, it was required to hold calcium sulphate in storage as a 75% aqueous, easily pumpable slurry of low viscosity.

Performance in these applications is generally assessed by the ability of a dispersing agent to allow preparation of the high-solids slurry with minimum addition level of dispersing agent to allow dispersion and minimum slurry viscosity at a given solids content.

To illustrate the differences in results one can obtain by merely altering the phosphating agent whilst standardising on the nonionic base, we have taken high-solids-content calcium sulphate dihydrate slurries with differing concentrations of dispersants and measured viscosity by Brookfield Viscometer (Table VI).

TABLE VI

Product	Phos-phating agents	Solids content	Viscosity in Centipoise: Concentration of Dispersant (% on dry pigment) 0.75	0.85	0.95	1.05	1.15	1.25	1.35	1.45
2-ethyl hexanol + 2 EO Phosphate	P_2O_5	71	3400	2850	1000	480	280	-	-	-
		74	4650	2500	1220	760	470	465	450	-
		77	6700	3200	1480	870	600	570	500	-
2-ethyl hexanol + 2 EO Phosphate	TPA	71	-	-	7800	4200	3150	2870	2600	2580
		74	-	-	-	-	8180	7750	6800	6470

Here, it is essential to use a P_2O_5-derived phosphate ester.

(b) Phosphate esters as hydrotropes

A hydrotrope is a compound that performs the function of increasing the solubility of another compound in a selected media whilst not affecting the inherent properties of that compound.

This property is particularly important in the formulation of highly built liquids for industrial-cleaning application where hydrotropes are preferred to hydrophilic surfactants.

The solubilisation of nonionic surfactants into high concentrations of electrolytes is a particularly difficult task, even more so if those nonionic surfactants are low foam in character based on copolymers of ethylene oxide and propylene oxide.

The choice of phosphate ester hydrotrope can be critical in this application which is illustrated by phosphate esters of phenol + 2,4 and 6 moles of ethylene oxide. For simplicity the ratio of nonionic to phosphorylating agent has been kept constant at 3:1.

The mole ratio of nonionic base to phosphating agent was varied between 2:1 - 4:1, with optimum performance in each case being given by the 3:1 ratio.

Tables VII and VIII illustrate the performance differences.

TABLE VII Solubilisation of 5% C_{12-15} alcohol + 3 moles EO into 18% w/v. Tetra-potassium pyrophosphate solution

Hydrotrope		Mole ratio	% Hydrotrope for clarity 20-60°C
Nonionic base	Phosphating Agent		
Phenol + 2 EO	TPA	3:1	12.5
	P_2O_5	3:1	15.0
Phenol + 4 EO	TPA	3:1	9.0
	P_2O_5	3:1	15.0
Phenol + 6 EO	TPA	3:1	8.5
	P_2O_5	3:1	15.0

TABLE VIII Solubilisation of 5% nonylphenol + 9 moles EO into 18% w/v Tetra-potassium pyrophosphate solution

Hydrotrope		Mole ratio	% Hydrotrope for clarity 20-60°C
Nonionic base	Phosphating Agent		
Phenol + 2 EO	TPA	3:1	5.0
	P_2O_5	3:1	15.0
Phenol + 4 EO	TPA	3:1	4.5
	P_2O_5	3:1	15.0
Phenol + 6 EO	TPA	3:1	4.0
	P_2O_5	3:1	15.0

In both cases, the tetraphosphoric acid-derived materials are superior to those derived from phosphorus pentoxide.

It is hoped that this example illustrates the need for an in-depth knowledge of the surfactants you use, in order to fully appreciate both positive and negative factors affecting performance.

The application areas cited in the various industry areas which are tabulated in this paper illustrate the diversity of surfactant usage. It is difficult to identify a single industry which does not directly or indirectly utilise surfactants, although the end-user may not always be familiar with the term 'surfactants'. In some cases, these materials are purchased as 'defoamers', 'dispersants', 'de-watering aids', etc., where the surfactant has become an effect or performance chemical.

To the surfactant raw-material producers, converters and formulators, the industrial applications of surfactants continue to provide stimulating business and development opportunities for those willing to dedicate the required techno-commercial skills and resources.

REFERENCES

(1) Prof.B.W. Werdelmann, 'Surfactants in our World-Today and Tomorrow', opening address at World Surfactants Congress, Munich, 6th-10th May 1984.

(2) 'Surfactants in Industrial Applications - Western Europe Forecast to 1990', Colin A. Houston Associates, August 1985.

(3) 'Surfactant Intermediates and Soap - World Balances 1978-83 with Forecasts to 1990', Colin A. Houston Associates, October 1982.

(4) Special Report, Manufacturing Chemist, 27 February 1986.

(5) 'Anionic Surfactants based on Phosphorus', T. Fay, G. P. Sheridan, D. R. Karsa (Diamond Shamrock Europe), Proceedings of 'XIIth Journadas de Comite Espanol de la Detergencia (CED/AID)', Barcelona, Spain, 27.02.81.

Specific Properties of Silicone Surfactants and Their Applications

By G. L. F. Schmidt
ZENTRALLABORATORIUM TH. GOLDSCHMIDT AG, GOLDSCHMIDTSTRASSE 100, D-4300 ESSEN 1, WEST GERMANY

1. Introduction

Because of their extraordinary properties silicone surfactants find applications in those areas where organic products are largely ineffective. This is primarily in organic media or in order to decrease the surface tension of aqueous systems to a greater degree than conventional surfactants. In addition to some new laboratory results, this paper is based mainly on four comprehensive publications,[1-4] which in themselves make reference to a broad range of previous literature.

2. Chemistry of Silicone Surfactants

2a) Structure

Silicone surfactants as to be discussed hereafter consist of a polydimethylsiloxane backbone in which some of the methyl groups are substituted by organic groups of hydrophilic or organophilic character, both frequently termed lyophilic. The silicone backbone itself behaves as the lyophobic part. It may be considered to be a block of 'super-hydrocarbon' because of the high concentration of methyl groups present. Irrespective of its form, whether stretched or helical, its siloxane groups are present at the surface resulting in poor compatibility with organic media. This illustrates one of the differences between polysiloxanes and pure organic entities. This has already been discussed critically by G. Koerner.[3] The structures of silicone surfactants are mainly of two types. One type bears its lyophilic groups at the end of siloxane chains, either branched polyfunctional (Figure 1) or linear difunctional (Figure 2).

The second type is a rake or comb structure bearing modifications grafted within the chain (Figure 3). If the rake lost all but one of its teeth a monofunctional molecule (Figure

4) would result. This is a normal feature of organic surfactants many of which are monofunctional. Silicone surfactants on the other hand, are primarily polyfunctional.

The symbol A in the structural formulae of Figures 1 to 4 can be hydrophilic groups of anionic, cationic, amphoteric, or non-ionic character or organophilic groups such as alkyl alkylaryl esters amides or polyethers (e.g. polypropylene oxide or polytetrahydrofuran) all of which have previously been synthesized.

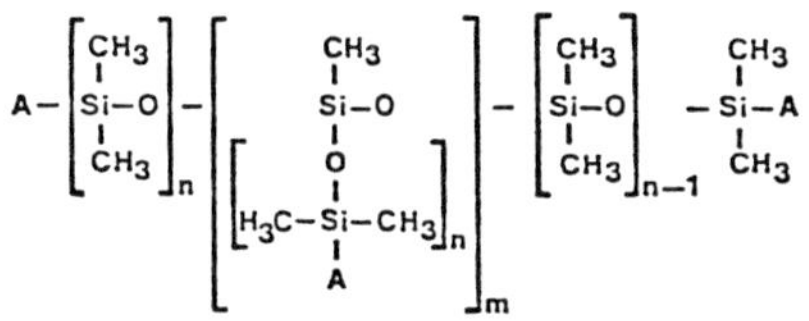

Figure 1 **polyfunctional branched type**

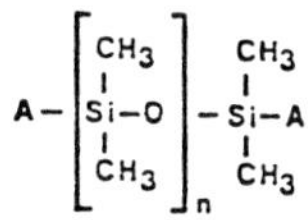

Figure 2 **difunctional linear type**

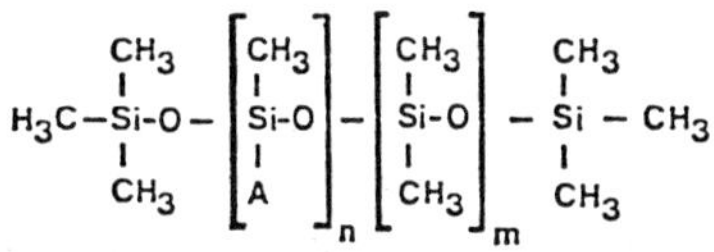

Figure 3 **polyfunctional linear type**

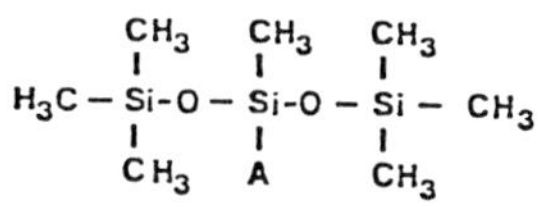

Figure 4 **monofunctional type**

2b) Synthesis

Normally the silicone backbone is pre-formed and then linked to organic molecules carrying the lyophilic function that is required in the finished product.

One route is to start from silicones which contain some SiCl groups. These react by condensation with COH of monoles bearing a lyophilic principle. The same monoles could of course, react with silicones containing $SiOCH_3$ or $SiOC_2H_5$ groups instead of chlorine in this case by transesterification. The linkage resulting from this type of reaction is an Si-O-C bond.

The other method of combination is to produce a silicone that contains, besides $SiCH_3$, some SiH groups within the chain. These react with olefinic double bonds of molecules carrying the lyophilic principle which is to be inserted into the required surfactant. The resulting Si-C bond is of higher chemical stability[3] in some circumstances.

Further reactions with the chemical functions linked to the silicone backbone may be carried out. One example is the variety of possibilities offered by an epoxy group that has been previously linked to the silicone.[4] It in turn can enable cationics to be produced by the addition of amines or anionics by HSO_3.

3. Physical Properties

3a) Surface activity

Silicone surfactants can also work in organic media, where many organic analogues can not. Figure 5 shows for example how a special polyether-silicone copolymer lowers the surface tension of an organic solvent, which contains a coating polymer, by about 4 mN m^{-1}. From Figure 6 it can be seen that a polyether-silicone copolymer lowers the surface tension of a polyether by more than 10 units.[2]

This property is already used in non-aqueous paints to achieve wetting, levelling, and pigment dispersion. Another well known application in organic media is to stabilize PUR foams during their fluid state of production.

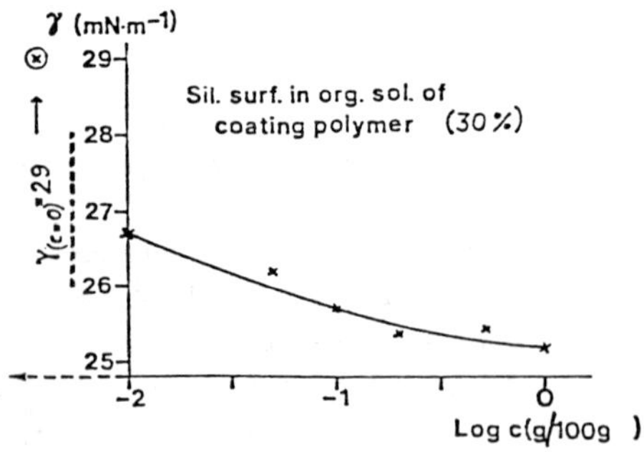

Figure 5

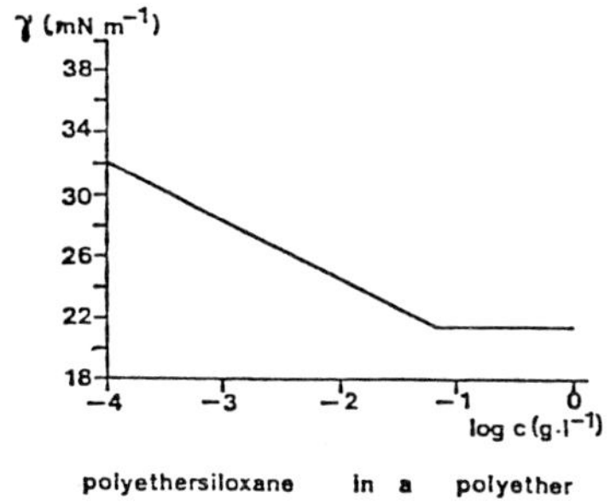

Figure 6

Silicone surfactants and organic surfactants differ to some extent quantitatively: silicones decrease the surface tension of water down to 20 mN m^{-1}, whereas organics reach only about 30 mN m^{-1} in most cases. Figure 7 illustrates the surface activity obtained by the anionic silicone of Figure 8 in water.

In order to complete the comparison with organics, similarities must also be pointed out. Critical micelle concentrations of surfactants depend on the chemical character of the modifying group. This determines the compatibility with the medium under investigation. In addition the surface tension achieved in a special medium depends on this property. The wetting angle on a particular solid substance is ruled by interface interactions.[2] Speed of wetting depends on the mobility of surfactant molecules, the smaller the faster.

As the silicone backbone is very flexible, most silicone surfactants remain fluid even at high molecular weights. There are exceptions when certain organic groups are attached. These are linear alkyl groups containing more than 17 C-atoms or polyethylene oxide blocks of molecular weights higher than 750. In this case solid products are only obtained at room temperature if the organic groups are present in a sufficient quantity per molecule. Therefore, under normal conditions, we cannot assume any crystalline monolayers of silicones on surfaces or in micelles. Regarding those micelles there is only limited knowledge about shape, size, or character. Their existence is only deduced from observations very likely correlating to them.

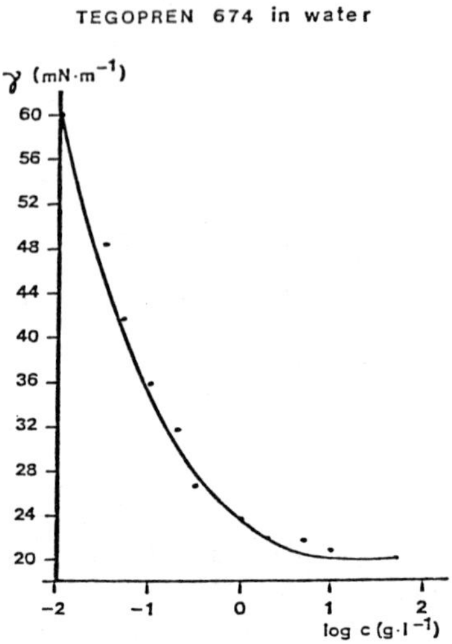

Figure 7

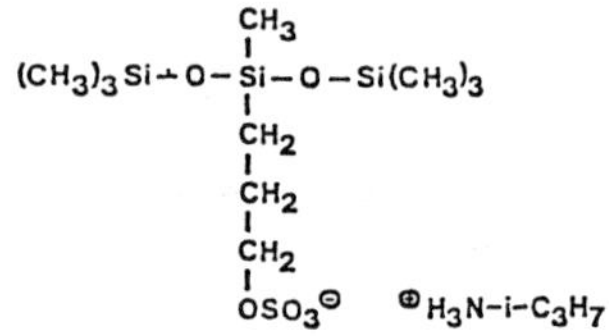

Figure 8

Finally, a similar feature of organic and silicone non-ionics is the property of their cloud point. Silicone polyethers can achieve cloud points between 3 and 120 °C.

4 Applications

4a) Emulsifiers and dispersants

Copolymers of silicone and polyether can produce o/w emulsions of silicone oils using substances corresponding to the formulae of Figures 1 and 3 containing about 30 to 50 Si atoms per molecule and 6 to 10 polyether blocks. One of the best emulsifiers observed was a substance with a cloud point of 70 °C.

When using an organic emulsifier for silicone oils the optimum HLB vlaue to produce stable o/w formulations is 13 ± 2. Thus we supposed this range of HLB values also for the optimal silicone non-ionic. Then as a countercheck we made experimental HLB determinations via emulsification of mineral oil (paraffinum liquidum). The same polyether siloxane was combined with an organic emulsifier of well defined HLB. From these experiments we obtained a figure of 24±2 for the HLB of the a/m product compared with an expected figure of 13±2.

This extreme discrepancy can be explained by the fact that in silicone oil emulsion the methyl siloxane moiety contributes effectively to the HLB value. In the case of mineral oil the silicone moiety is neutral and only the ration of EO to PO determines the emulsifying quality of the same copolymer.[6]

4b) Wetting and adhesion

For the adherence of binders as well as coatings, wetting interaction with the substrate is necessary. One case in which this problem was not satisfactorily resolved will be explained here.

During industrial production of clothes the use of thermoplastic adhesives to join fabrics has become very common besides sewing. However, raincoat fabrics, which are made waterproof by silicones, are not only repellent against water but also against bonding thermoplasts, e.g. PA The adhesive strength is reduced five times compared with non-repellent fabrics. If, however, small amounts of a surface active silicone are added the contact angle of the molten thermoplast is lowered and in fact bonding strength increases. Although it only reaches about half the original value the result is advantageous for application. First experiments have been performed by coating PA powder with fluid silicone surfactants. The treated powder was then sieved onto a clean fabric, melted, and cured by infra-red light Water-repellent fabric and an untreated one respectively were laminated with the adhesive using a flat iron. Bonding forces were measured using a spring balance. The results are shown in Table 1.

The most effective organophilically modified silicone surfactants for thermoplastic substrates mentioned were dimethylsiloxanes of medium molecular weight (1000 - 3000), in which some of the methyl groups were substituted by ester, $Si(CH_2)_3$-O-C(O)R or nonylphenoxy groups, Si-O-C_6H_4-C_9H_{19} (see Figure 9). Other substrates may require other modifying groups to achieve compatibility. We have indications that for example, polyether-silicone copolymers are effective in bonding agents of

Table 1 Fabric bonding

fabric (cotton)	binder(PA)	adhesive strength(N)
untreated	neat	12.5
siliconized	neat	2.5
Siliconized	with silicone surfactant	6.5

higher polarity such as polyvinylalcohol. Our laboratory is able to tailor chemical modifications to meet specific requirements.

We should mention here that the picture is not quite as simple as we have illustrated and a lot of further work needs to be done to clarify how wetting and sticking properties correlate.

4c) Tribological findings

With regard to lubricants wetting alone is not the only requirement: speed of wetting is very important. This is necessary where films of an oil have to spread readily in a continuous or periodical way at high speed. One example is the extrusion and further processing of thermoplastic fibres. During their drawing, winding texturizing, and thermosetting, such lubricants are used. The spin finish may be a normal silicone oil a chemically modified one or a blend containing the latter at high or low concentration.

Surface activity is used here as a bulk property or in an organic fluid in order to lower its surface tension. How polyether siloxanes perform in a polyether that consists of PO and EO units randomly distributed is shown in Figure 6. Ester oils and even mineral oils can be influenced in their surface tension An adequate modification is illustrated in Figure 9.

With this compound the surface tension of a mineral oil was decreased from 27.2 mN m^{-1} down to 24.2 mN m^{-1}. This change may not seem great yet it may break a wetting barrier on very non-polar surfaces, e g. polypropylene.

At higher dosage than usual for surfactants, bulk properties of the modified silicones appear and influence the lubricant's

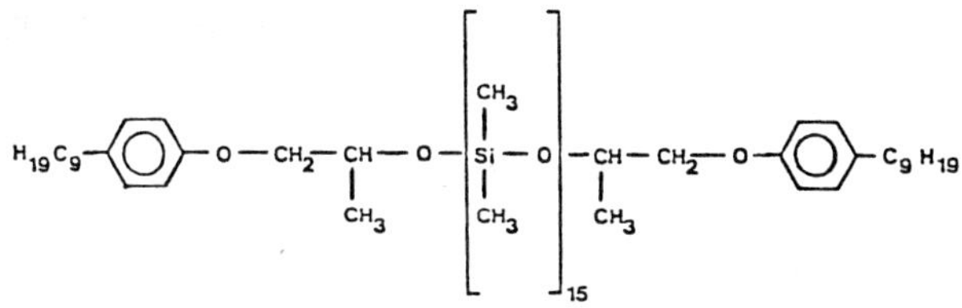

Figure 9 TEGOPREN 1038

behaviour. Thermostability and good lubricity at low pressure occur as well as their parting properties which doubtless correlate with the lowering of friction value.

4d) Release agents

The application of silicone surfactants in parting agents is illustrated briefly by one example.

For demoulding of concrete, oily or aqueous systems are used as auxiliaries. Because of their limited spreading properties, irregular coating of moulds may occur. Small amounts of silicone surfactants can overcome this disadvantage significantly. This can be shown by the excellent reproduction of the sharp-cornered edges of moulds or the grain of wooden boards. During the release process no dust is generated, which is a further advantage

The organophilic modification choosen for oily systems was polypropylene oxide, whereas the hydrophilic one for aqueous systems was polyethylene oxide.

5. Conclusion

Further investigations which are not reported here especially concerning wetting and more complex molecular structures are in progress. However, the nearly unlimited chemical variability of silicone surfactants should encourage experts to proceed in examining this class of substances.

References

(1) G. Koerner, Seifen-Ole-Fette-Wachse, 1972, 98, 335.

(2) G. Koerner, G. Rossmy and G. Sänger, Goldschmidt Informiert, 1974 Nr. 29.

(3) G. Koerner, Goldschmidt Informiert, 1982, Nr. 56 (English Edition), 2.

(4) J.J. Kollmeier and R.D.Langenhagen, Goldschmidt Informiert, 1984 Nr. 63 (English Edition), 41.

(5) A.W Neumann and D. Renzow, Z. Phys. Chem. (Neue Folge), 1969, 66, 11.

(6) S C. Vick, Soap/Cosmetics/Chemical Specialities, May 1984, 36 ff.

Surfactants Widely Used in the Concrete Industry

By B. El-Jazairi
FEB (GREAT BRITAIN) LIMITED, ALBANY HOUSE, SWINTON HALL ROAD, SWINTON, MANCHESTER M27 IDT, UK

1. Introduction

Although the term surfactant is not commonly known or used in the concrete industry, a large proportion of everyday concrete produced on site or by the precast and ready mixed concrete industry contains through the use of admixtures, one or more surface active agents. Surfactants, in combination with other chemicals or materials, are exclusively used by the construction and building chemical industry in formulating and manufacturing the different types of concrete admixtures. The use of surfactants in concrete is, therefore, closely associated with the use of concrete admixtures and their usage has increased considerably in recent years.

Well over 150 proprietary admixtures are available in the U.K. market and it is estimated that some 30 to 40 percent of the concrete produced in this country contains at least one admixture. In 1970 and 1975 some 20.8 and 24.7 thousand metric tons were produced in the U.K. respectively. It is estimated that the production in 1985 reached over 35 thousand metric tonnes, a third of which may be related to admixtures where surfactant forms the main active component. Worldwide over 700 patents related to admixtures, many containing surfactants, were registered in the last 10 years.

These figures give a clear indication of the increasing rate at which admixtures, and by association surfactants, are being used in concrete.

2. Outline of the Nature of Concrete and the Cement Paste

Portland cement concrete is a composite material composed of a "chemically" inert mineral filler (the aggregate), embedded in a hard matrix of hydrated cement that fills the space between the aggregate particles and binds them together. The main components for the making of concrete are:

- Mineral fillers
- Cement
- Water

The mineral fillers are granular materials derived, for the most part, from natural rock, natural gravel, crushed stones and sand. The mineral fillers usually occupy about 70%-80% of the volume of concrete and can, therefore, be expected to have an influence on its plastic and hardened properties.

The cement and water are the chemically active components. When Portland cement is mixed with water it forms a paste and its constituent minerals (tricalcium silicate, dicalcium silicate, tricalcium aluminate, the ferrite phase and gypsum) undergo a series of chemical reactions, which are responsible for the eventual hardening of concrete. This is the process of HYDRATION, and the new solids formed on hydration are collectively referred to as HYDRATION PRODUCTS.

When water is added to concrete mix, the cement-paste matrix remains plastic for a short time (few hours). The paste during its plastic state determines the rheological properties - THE WORKABILITY of FRESH CONCRETE. As the reaction between cement and water continues, the mass of the paste loses its plasticity and the fresh concrete loses its workability.

This early period in the hardening process, which is associated with the transition of concrete from the plastic to solid state, is called the 'SETTING TIME', although there is no well defined break in the hardening process. The hardening process and hydration process usually continues at a decreasing rate for months and sometimes for years. The hardened cement matrix bonds the mineral fillers together in concrete and as such it is the main bearer of the physical and chemical properties of HARDENED CONCRETE.

The water in concrete mix has essentially two functions:

i. to react chemically with cement to provide a strong, hard material;

ii. to make the concrete sufficiently workable for it to be easily placed and compacted.

To obtain workable concrete requires almost double the amount of water necessary for the complete chemical reaction of cement. The excess water, whilst necessary to produce workable concrete, will gradually evaporate as the concrete hardens. This evaporation leaves voids or pores in the concrete. The resulting porosity will reduce the strength of concrete and will also reduce the resistance of concrete to the ingress of any harmful materials. The more water there is in a mix, the weaker will be the concrete. For this reason the amount of water added to a concrete mix, usually expressed as the WATER/CEMENT RATIO (W/C), is considered the main factor affecting the porosity and strength of concrete and therefore its long term durability.

Most ordinary cements nowadays are ground to a fine powder having a surface area in the range of 300-500m^2 kg^{-1} and particle size distribution ranging from 0.5 to 70μm. At the initial stage, when water is added to cement, the aqueous phase immediately becomes highly alkaline (pH 12-13).

Through solubilities and hydration reactions of the various constituents, the aqueous phase contains different electrolytes[1]; viz: (SO_4^{2-} Ca^{2+}, OH^-, Na^+, K^+ and traces of Al/Fe/Si ions). The concentration of the different electrolytes varies with time but the aqueous phase remains supersaturate with calcium hydroxide throughout. It is therefore essential, when considering the use of surfactants in concrete, to ensure that they are compatible and stable in highly alkaline environments and remain effective in the presence of high electrolyte concentrations.

3. CONCRETE ADMIXTURES

3.1 Definition

An admixture is defined[2] as a proprietary mixture of chemicals which is added in small quantities (less than 5% by weight of cement) as an ingredient to a batch of concrete immediately before or during mixing in order to modify or impart certain properties of concrete in their plastic and or hardened state, or their passage from fresh to hardened state.

3.2 Classification

The large number of products marketed as admixtures can be basically grouped into the following categories[2]:

3.2.1. Chemical Admixtures

These are based on water-soluble chemical compounds that are primarily added either to control the setting time and early hardening of fresh concrete and/or to reduce its water requirements (accelerators, retarders, plasticisers and superplasticisers); or to entrain air to improve the frost resistance of concrete (air entrainers).

3.2.2. <u>Miscellaneous Chemical Admixtures</u>

These include all those materials that do not come under one of the foregoing categories, but may have been developed for special application. Included in this category amongst others are, gas forming, grouting, expansion producing, bonding, colouring, floculating, fungicidal, germicidal, insecticidal, damp-proofing, permeability reducing, corrosion inhibitors and polymers.

3.3.3. <u>Mineral Admixtures</u>

These are finely divided solids, added to concrete to improve its workability, to provide additional cementing properties and to improve durability. PFA, slag and silica fume are examples of these admixtures.

In many of the admixture formulations in all three categories, surfactants are used either:

i. as the main functional constituent which has direct action in modifying certain properties of concrete: surfactant based admixtures, or

ii as a minor constituent for an admixture based on chemicals and materials other than surfactant. In this case the addition is to improve wetting, spreading, emulsification, dispersion and stability of the admixture formulation itself or to enhance such properties of the admixture in concrete.

In all cases, whether the surfactant in an admixture formulation is the major or the minor constituent, it will affect physical and physico-chemical actions in the cement-water system, such as surface tension, adsorption, thixotrophy, foaming, colloidal state, deflocculating effects, and rate of reaction between cement and water during setting and hardening.

For these reasons it is essential, before the use of any surfactant in an admixture formulation, to identify its short and long term effect on concrete.

4 SURFACTANT BASED ADMIXTURES

4.1 WATER REDUCING ADMIXTURES

4.11. Definition

Water-reducing admixtures (commonly known as plasticisers) are water soluble organic compounds which lower the water required to attain a given workability in plastic concrete without significantly affecting the air content or alternatively achieve a greater workability without increasing the water-cement ratio. The water-reducing admixtures are classified in three types covering both their primary water-reducing functions and their secondary functions:

i Normal water-reducing which does not significantly affect the setting time.

ii Accelerating/water-reducing where the secondary function decreases the setting time and may act as an accelerator.

iii Retarding/water-reducing where the secondary function prolongs the setting time and may act as a retarder.

The normal water-reducing admixtures are based either on purified lignosulphate, which usually have a low sugar content (expressed as reducing sugars content); or on using low dosage of hydroxycarboxylic acid or polysaccharides type admixtures.

The accelerating water-reducing admixtures invariably contain accelerating chemicals such as calcium chloride, triethanolamin or calcium formate, in their formulation.

The retardation effect of retarding water-reducing admixtures is almost without exception due to the chemical nature of the surfactant.

Water-reducing admixtures are anionic surfactants. They are usually divided into three principal categories based on the general composition of their active ingredients:

i Salts and derivatives of lignosulphonic acids. (e.g. Calcium and or sodium lignosulphonates)

ii Salts and derivatives of hydroxycarboxylic acids. (e.g. sodium sats of gluconic, tartaric, citric acids)

iii Low molecular weight polysaccharides. (e.g. glucose, sucrose)

4.12 Mechanism

Cement particles in water carry residual charges on their surfaces, which may be positive or negative, or both. In cement paste, as a result of opposing charges in the cement grains, the cement particles can exert electrostatic attractions, causing the particles to flocculate. The resulting agglomerations incorporate considerable amounts of water adsorbed in the solid surfaces and in the agglomerate inter-spaces, leaving less water available to reduce the viscosity of the paste and hence increase the workability of concrete.

The addition of water-reducing admixtures as a surfactant are primarily adsorbed onto the cement particles at the solid-water interface. The adsorbtion neutralises the inter-particle attraction and imparts a negative surface charge uniformly distributed on the particle surfaces. This adsorbtion results in the breaking down of the agglomerates and produces a fully dispersed system in the cement paste, which affects the reduction of the viscosity of the paste, and hence the increase of workability of concrete.

4.13 Application

Water reducing admixtures can be used in the following ways:

i To increase the workability of concrete mixes without the addition of extra water. This enables easier placing and compaction of concrete on-site, especially in congested areas, without affecting its strength.

ii To increase the early and ultimate strength of concrete by maintaining the same workability of concrete mixes but at a reduced water-cement ratio. Water reduction of 8-12% can be made by water-reducing admixtures.

4.2 SUPERPLASTICISERS

4.21 Definition

Superplasticisers are a new class of high performance, water-reducing admixtures. They can be defined as chemicals or mixtures of chemicals which, when added to concrete, can either achieve a water reduction of 15 to 30% or impart extreme workability.

Superplasticisers are anionic surface active agents. They are linear polymers containing a sulphonic acid group attached to the polymer backbone at regular intervals. Two main polymer types are the basis of most superplasticiser formulations used in the U.K.:

i salts of a sulphonated melamine formaldehyde condensate

ii salts of a sulphonated naphthalene formaldehyde condensate.

4.22 Mechanism

The mechanism at which the superplasticiser molecules interact with cement at the solid-water interface is similar to the mechanism of interaction of normal water-reducing agents. The difference is that the electrokinetic potential (zeta potential) resulting from the adsorbtion of superplasticiser molecules is much higher. Thus it brings a higher negative charge at the cement/water interface, resulting in a higher repulsion force between the particles in the system.

4.23 Application

Basically, superplasticisers are used to produce:

i flowing concrete

ii high-strength concrete

Flowing concrete: superplasticisers can be used to advantage in areas of difficult placement, such as congested reinforcement or in placements where adequate consolidation and vibration cannot easily be achieved. Due to its almost self-levelling flowability, flowing concrete is used widely for placing floor and pavement slabs with little or no vibration. Superplasticisers are also widely used for pumping concrete. In all these types of application care should be taken in mix design to ensure segregation will not occur. Increasing the proportion of fines by 5% usually counteracts the segregation problem.

High-strength concrete superplasticisers are capable of reducing the amount of concrete by 20-30%, whilst maintaining the same workability. This results in a substantial increase in both the early and ultimate strength of concrete.

For this reason, superplasticisers can achieve high early strength and are widely used as alternatives to accelerators. They are used for early demoulding in the construction and precast industries and also in most applications where accelerators are normally used.

4.3 AIR-ENTRAINING ADMIXTURES

4.31 Definition

Air-entraining admixtures are organic surface active agents which entrain a controlled amount of air in concrete, in the form of a stable dispersion of uniformly distributed air bubbles, ranging in size between 20 to 1000 microns in diameter.

The principal surfactant used in most commercial products are formulated from the following categories of materials:

i Sodium salts of wood resins (neutralised vinsol resin). This is the most widely used surfactant, which produces the most stable and efficient small air bubbles dispersion in concrete.

ii Sodium salts of alkyl aryl sulphonates with alkyl residue of about C_{12}. This is a common class of surfactants, e.g. sodium p-dodecyl benzene sulphonate. This category produces less stable air dispersion in concrete and often produces large air bubbles.

iii Salts of unrefined lignosulphonates. These are poor air entrainers and are mostly used to produce air entraining/water-reducing admixtures. Sometimes they are blended with one of the above agents to boost its air entraining capacity.

iv Salts of proteinaceous materials derived from the animal and hide processing industries. These consist of salts of a complex mixture of carboxylic and amino acids. They produce larger air voids and few admixtures are of this type.

4.32 Mechanism

Effective air-entraining admixtures are anionic surface active agents. They are water-soluble salts of organic compounds. The molecules of the compounds consist of a non-polar hydrocarbon chain or some other hydrophobic (water-repelling) group, joined to a polar or hydrophillic (water-attracting) group[3] such as $R\text{-}COO^-$, $R\text{-}SO_4^{--}$ or $R\text{-}NH_3^+$. The polar group makes the molecule soluble in water at an appropriate temperature, whilst the non-polar group tends to be expelled from the water. As a result of these two opposing tendencies, surface active molecules tend to concentrate at interfaces between different phases.

At air-water interface the majority of the polar groups are oriented towards the water phase and the majority of non-polar groups are oriented towards the air phase .

This phenomenon, known as 'non-specific adsorbtion', lowers the surface tension of the water and promotes bubble formation. The air bubbles formed become negatively charged at the water-air interface and the negative charge is directed towards the water phase. This property counteracts the tendency for the dispersed bubbles to coalesce through the action of electrokinetic repulsive forces.

At the solid-water interface, the cement particles are positively charged, at least at the initial stage of solid contact with water. Here the polar (hydrophillic) part of the surface active molecules are adsorbed onto the cement particle surface and the non-polar (hydrophobic) part is directed towards the water.

This adsorbtion makes the cement surface hydrophobic, so that air may displace water from it and, as a consequence, air bubbles may adhere to the solid.

The adhesion of bubbles to hydrophobic cement particles is an important factor in the air-entrainment process. The adhesion phenomenon promotes the size reduction of air-entrained bubbles by the partition process of large bubbles through the multi-directional flow of cement particles, ensures uniform distribution of air bubbles throughout the concrete mix and stabilises air content in concrete. In contrast, hardened, plain concrete (without an air-entraining admixture) contains a certain amount of irregularly distributed air voids.

Hardened concrete with air-entraining admixtures, unlike plain concrete, entrains different types of air voids. Here air voids are more regular, ranging in diameter from 0.01 to 1.005mm, the major part ranging from 0.025 to 0.250mm. These air voids are minute, discrete, uniformly distributed air bubbles. The distance between individual air voids is usually between 0.1 - 0.2mm. The size and regular distribution of the entrained air provides spaces where forces which would cause disintegration can be dissipated. The air bubbles in air-entrained concrete also disrupt the continuous capillary voids and thus affect the reduction in permeability. In other words air-entraining admixtures produces concrete with stable dispersion of small air voids, both in the plastic and hardened states.

4.33 Effect on Properties of Concrete

The effect of air-entraining admixtures (AEA) is dependent on dosage, cement type, aggregate type and grading, mix proportions, workability, temperature, type of mixer and mixing time.

The plastic mix becomes more cohesive with AEA and this property is of particular value where sands and aggregates with poor grading are used.

Bleeding and segregation is substantially reduced by the use of air entraining admixtures. For a concrete of medium workability a 5% air-entrainment may half the amount of bleeding of similar concrete without AEA. This property is one of the major advantages for the use of this admixture.

Air entrainment in concrete will reduce the strength. 1% of air entrainment may reduce the strength by almost 5.5% In practice the reduction of strength caused by air-entraining admixtures is compensated for by reducing the water-cement ratio.

4.34 Durability

The increased air-entrainment of concrete from 3% to 6% considerably increases the resistance of concrete to the disintegrating action of freezing and thawing.

The disruption of the continuous capillary, the reduction in segregation and bleeding and the greater cohesiveness caused by the air entrainment, tend to produce a more durable, watertight concrete.

4.35 Application

The main application of air-entraining admixtures (AEA) is their use on concrete exposed to the action of frost and de-icing salts. The production of air entrained concrete of 4.5% $\pm$ 1.5% is specified by the Department of Transport for Road and Works, for concrete paving for roads and airfields, to improve its resistance to these actions. Other applications of AEA are to compensate for a poor aggregate, which produces harsh mixes and causes bleeding. It is also widely used for pumping and extruded concrete.

5 Conclusion

In conclusion the surfactant based admixtures, described in this paper, are usually used to modify the following properties in concrete:

Fresh Concrete (Plastic state)

- To increase the workability without increasing the water/cement ratio.
- To improve the cohesion
- To reduce bleeding and/or segregation (separation of the individual phases).
- To retard the setting (stiffening) process

Hardened Concrete

- To increase the rate of early strength development.
- To increase the ultimate strength.
- To decrease permeability and improve durability.
- To increase the resistance to frost action.

This paper was confined to the principal surfactant based Admixtures. These surfactants account for the largest volume of organic materials and probably over 90% of all types of surfactants used in concrete. Because of their wide use, they are well established in terms of chemistry, mode of action and their effect on the short and long term properties of concrete. Since there is a lack of specific information on the effect of other types of surfactants on concrete, their use at the present time could be said to be based more on art than science.

References:

1. D.D. Double Philosophical Transactions of the Royal Society of London,1983, series A, 310, (No.1511) pp 53-65.

2. ACI Report 'Guide for Use of Admixtures' No. 68-56.

3. P.C. Kreiger Proceedings of the International Congress on Admixtures, The Construction Press, London, 1980, pp 1-16.

Antifoams in Gas–Oil Separation

By I. C. Callaghan, S. A. Hickman, F. T. Lawrence,* and P. M. Melton
BP RESEARCH CENTRE, CHERTSEY ROAD, SUNBURY-ON-THAMES, MIDDLESEX TW16 7LN, UK

1. INTRODUCTION

In its natural state crude oil contains dissolved gases (typically 10% by weight) which are held by high reservoir pressure. When this "live" crude oil is extracted and passed into the low pressure environment of a gas-oil separator the dissolved gases are liberated, ideally producing a gaseous phase, comprising natural gas, which is taken from the top of the separator, and a liquid phase, comprising "stabilised" crude oil, which is taken from the bottom (Figure 1 (a)). Often, however, the disengagement of gas from crude oil is not clean, and a foam is formed (Figure 1 (b)). This may lead to the loss of crude oil *via* the gas stream and damage to downstream gas processing equipment[1]. There are three ways in which this wasteful process can be avoided. Firstly, prior knowledge of the foaming properties of a crude oil may allow the design of a separator large enough to cope with foam formation. Secondly, the amount of foam can be reduced by the injection of an antifoam or, thirdly, the gas-oil separator can be equipped with a physical means of destroying or preventing foam.

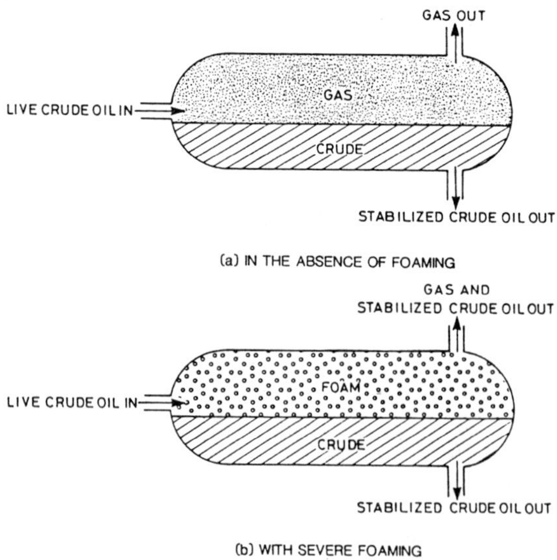

FIG 1 OPERATION OF A GAS-OIL SEPARATOR

The first method of avoiding crude oil carry-over has been considered in a paper which describes the application of the Optical Transmission Foam Meter[2](OTFM) to relate aqueous and non-aqueous foam collapse to filler gas solubility and liquid viscosity[3]. This study was undertaken because the ability to predict the foaming properties of a live crude oil from the physical properties of the gaseous and liquid components would facilitate gas-oil separator design. However, although several advances have been made in this respect (for example, crude oil foam stability increases with liquid viscosity[4]. Also, Sharovarnikov *et al*[5] have noted that the rate of aqueous foam collapse *via* interbubble gas diffusion is proportional to the product of the solubility of the filler gas in the liquid phase (water in their experiments) and the diffusion coefficient of the filler gas through the liquid phase), the ability to predict complete foam collapse profiles is still an important (but unattained) goal.

The foam collapse profiles of model aqueous and non-aqueous surfactant solutions were obtained using the OTFM which is shown in Figure 2. The OTFM measures the voltage output by the solar cells as the foam in the foam tube collapses (the exposed area-output voltage relationship for the solar cells is linear). The voltage-time profiles obtained in this way were converted into foam quantity-time profiles from knowledge of a baseline voltage corresponding to complete foam collapse. In the light of a previous investigation[6] the foam quantity-time profiles were fitted to the following equation, which comprises two exponential terms, using an iterative method[7]:

$$H_t - H_\infty = (H_o - H_\infty)\,(Z \exp[-K_1 t] + [1 - Z] \exp[-K_2 t]) \quad .. \quad (1)$$

where H_t = foam quantity at time, t (V)

H_∞ = final foam quantity* (V)

H_o = initial foam quantity (V)

Z = fraction of foam collapsing *via* mechanism (1)

$(1 - Z)$ = fraction of foam collapsing *via* mechanism (2)

K_1 = a rate constant giving the speed of mechanism (1) (S^{-1})

K_2 = a rate constant giving the speed of mechanism (2) (S^{-1})

Notes:

*It was necessary to include H_∞ in the equation because none of the foams collapsed completely (ie, $H_\infty \neq 0$).

Initial experimental results indicated that it was reasonable to use Equation (1) to describe foam collapse profiles. Further work suggested that K_1 and K_2 depended upon the composition of the filler gas and the viscosity of the liquid phase. It was concluded that an equation with the form of Equation (1) might provide a basis for the prediction of the foam collapse profiles of aqueous and non-aqueous systems and could ultimately be applied to crude oil foams.

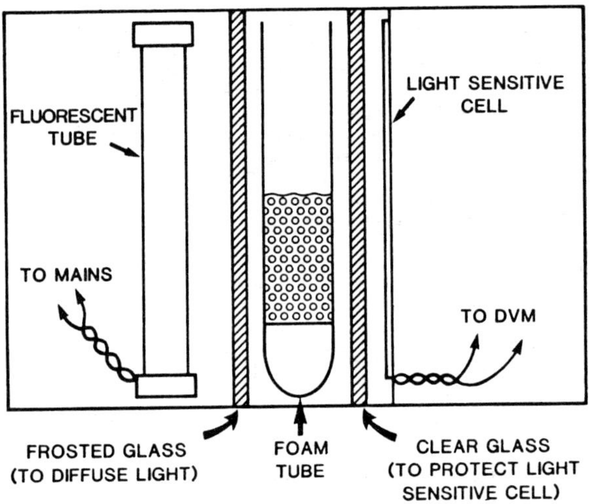

FIG2 THE OPTICAL TRANSMISSION FOAM METER

The second method of avoiding crude oil carry-over, the injection of an antifoam, will be considered in this paper which describes a study of the effect of crude oil composition upon antifoam ability. The investigation was undertaken in the light of experiments with the apparatus shown in Figure 3. In this technique simulated live crude oil is produced by mixing stabilised crude oil with liquid propane and butane gases at 4bar. The pressure is then relieved and the liquid gases boil and cause the crude oil to foam. When this method was used to asses the power of two antifoam types, a silicone glycol and an unsubstituted polydimethylsiloxane (PDMS) with aged crude oil both antifoams worked well. With fresh crude oil, however, only the silicone glycol antifoam worked. This observation was rationalised by suggesting that the (non-polar) unsubstituted PDMS had been solubilised (and rendered inactive) in the extra light hydrocarbon in the fresh crude oil while the silicone glycol (which has some polar nature) had been left unsolubilised.

An understanding of the factors which inhibit antifoam power under these conditions is of great importance since it relates to how well an antifoam will perform with pipeline crude oil, which has associated light hydrocarbon, relative to its performance in laboratory tests with stabilised crude oil. It will also indicate the range of crude oil compositions with which different antifoam types will work. Therefore, the "solubilisation theory" has been examined using a wide range of antifoams and crude oil/solvent mixtures.

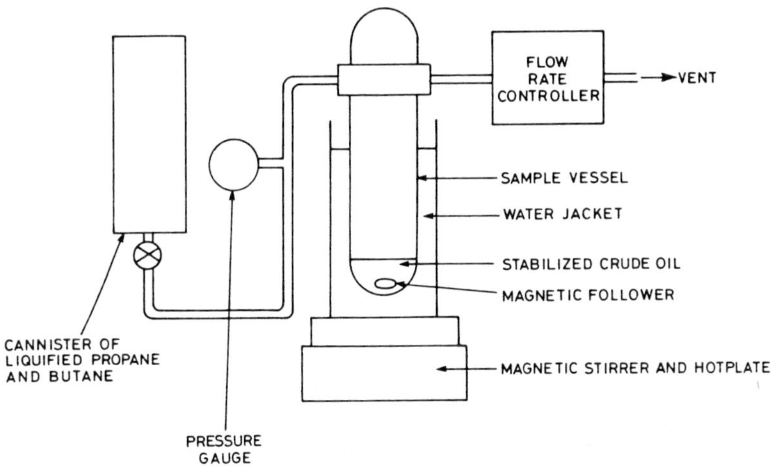

FIG 3 A PORTABLE FOAM APPARATUS

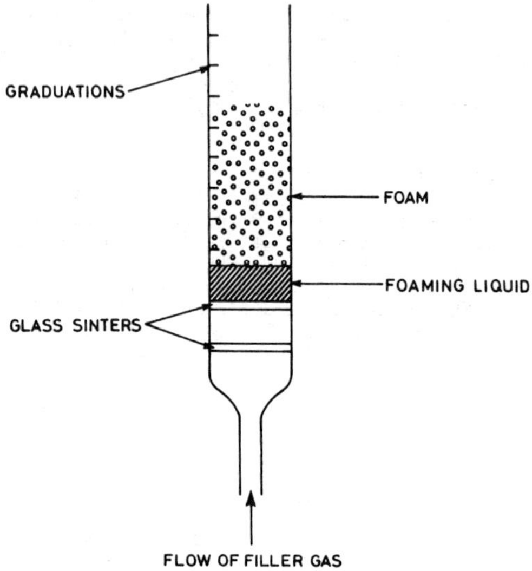

FIG 4 A GAS SPARGING TUBE

2. EXPERIMENTAL TECHNIQUE AND RESULTS

(a) The power of five antifoam types (at 5 (active) ppm) was measured with stabilised Magnus crude oil doped with 0, 10, 20, 30 and 40% of five different solvents using the gas sparging tube apparatus[4] shown in Figure 4. All determinations were at room temperature with a sample size of 2 cm^3 and a filler gas (natural gas) flow rate of $49 cm^3 min^{-1}$. Antifoam power was calculated from the following relationship:

$$AF\ (Xppm) = \frac{FH\ (0ppm) - FH\ (Xppm)}{FH\ (0ppm)}$$

where AF (Xppm) = the "antifoaming index" of antifoam at Xppm
FH (0ppm) = foam height given by "pure" crude oil
FH (Xppm) = foam height given by crude oil with Xppm antifoam

Hence, if AF (Xppm) = 1 the antifoam shows complete antifoaming action
= 0 the antifoam shows no antifoaming action
< 0 the antifoam shows profoaming action.

Antifoaming indices are shown in Table (1).

TABLE 1

THE EFFECT OF CRUDE OIL COMPOSITION UPON ANTIFOAMING INDICES

% SOLVENT		0*	10	20	30	40
Solvent	Antifoam**					
Hexane	1	0.74	0.44	-0.08	-0.20	-0.89
	2	0.78	0.57	0.02	-0.05	-0.59
	3	0.84	0.68	0.16	-0.05	-0.50
	4	0.88	0.84	0.84	0.78	0.66
	5	0.85	0.70	0.44	0.05	0.31
Octane	1	0.74	0.52	-0.18	-0.53	-0.33
	2	0.78	0.64	-0.03	-0.18	-0.33
	3	0.84	0.75	0.07	-0.10	-0.44
	4	0.88	0.84	0.78	0.80	0.76
	5	0.85	0.81	0.65	0.67	0.47
Stoddard solvent	1	0.74	0.85	0.00	-0.35	-0.81
	2	0.78	0.87	0.39	-0.09	-0.79
	3	0.84	0.87	0.58	0.06	-0.54
	4	0.88	0.94	0.86	0.74	0.72
	5	0.85	0.83	0.73	0.63	0.40
Toluene	1	0.74	0.73	-0.39	-0.55	-0.49
	2	0.78	0.77	-0.02	-0.33	-0.24
	3	0.84	0.83	0.27	-0.07	-0.17
	4	0.88	0.90	0.79	0.75	0.79
	5	0.85	0.85	0.67	0.64	0.37
Arklone'P'	1	0.74	0.51	-0.54	-0.20	-0.36
	2	0.78	0.73	-0.15	-0.06	-0.26
	3	0.84	0.75	0.19	-0.05	-0.21
	4	0.88	0.81	0.79	0.92	0.86
	5	0.85	0.68	0.59	0.52	0.38

* "Base" crude oil = stabilised Magnus
** Antifoam No. 1 = 12 500 cSt PDMS
2 = 60 000 cSt PDMS
3 = 1000 000 cSt PDMS
4 = fluorosilicone oil (Th. Goldschmidt's C 2610)
5 = silicone glycol (BPCI's AF 2001)

(b) The fall-off in the power of each antifoam with each solvent was calculated from the following equation:

$$\text{"fall-off index"} = \sum_{\%\ \text{solvent} = 10}^{\%\ \text{solvent} = 40} (AF_o - AF_s)$$

where AF_o = antifoaming index in absence of solvent
AF_s = antifoaming index in presence of solvent

Fall-off indices are shown in Table (2).

TABLE 2

ANTIFOAM POWER FALL-OFF INDICES

Antifoam/Solvent	1	2	3	4	5
Hexane	3.69	3.17	2.52	0.40	1.90
Octane	3.48	3.02	<u>3.08</u>	0.34	0.80
Stoddard solvent	3.27	2.74	2.39	0.26	0.81
Toluene	3.66	2.94	2.50	0.29	0.87
Arklone 'P'	3.55	2.86	2.68	0.14	1.23

(c) The solubilities of three antifoams in all the solvents were estimated by adding 8 cm^3 of each solvent to 2g of each antifoam in a 20 cm^3 sample bottle and agitating it. The sample bottles were then inspected immediately and after two weeks.

The results of these experiments are shown in Table (3).

TABLE 3

ANTIFOAM SOLUBILITY

(a) Immediately After Agitation Of Sample Bottle:

Antifoam:	1	2	5
Solvent:			
Hexane	1	2	3
Octane	1	2	3
Stoddard solvent	1	2	3
Toluene	1	2	1
Arklone'P'	1	2	1

(b) After 2 Weeks

Antifoam:	1	2	5
Solvent:			
Hexane	1	1	3
Octane	1	1	3
Stoddard solvent	1	1	3
Toluene	1	1	1
Arklone 'P'	1	1	1

1 : Single phase formed
2 : Antifoam - solvent boundary visible only on "up-ending" sample tubes
3 : 2 phases clearly visible

3. DISCUSSION

In general, the data in Tables 2 and 3 support the hypothesis that antifoam ability falls as the antifoam becomes more soluble in the crude oil.

(i) The three PDMS antifoams showed impaired ability with crude oil doped with all solvents. 12 500 cSt PDMS and 60 000 cSt PDMS are soluble in these solvents. It is probable that 1 000 000 cSt PDMS is too.

(ii) The results for the PDMS antifoams and crude oil doped with hexane, octane and Stoddard solvent indicate that antifoam power falls more with light solvent than with heavy solvent and more with low viscosity PDMS than with high (one exception, though, is the antifoaming power of 1 000 000 cSt PDMS with crude oil doped with octane).

(iii) The antifoaming power of high viscosity PDMS was less sensitive to added toluene and Arklone 'P' than was that of low viscosity PDMS.

However, the hypothesis that antifoam power falls as the antifoam becomes more soluble in the crude oil cannot account for all of the experimental results. The antifoaming ability of the fluorosilicone oil was not affected by the addition of the solvents and that of the silicone glycol was only slightly affected. This was in spite of the following:

(i) The fluorosilicone oil is soluble in Arklone 'P' (it was not possible to observe this with 100% antifoam. However, the 10% solution used in this study was completely miscible with Arklone 'P').

(ii) The silicone glycol dissolves in toluene and Arklone 'P'.

These observations suggest that the fluorosilicone oil and the silicone glycol do not need to be immiscible with the crude oil to show successful antifoaming action and so may not operate *via* the commonly accepted (droplet spreading) mechanism[8,9]

The results which show that the antifoaming abilities of the fluorosilicone oil and the silicone glycol are not greatly affected by the presence of hexane and octane in the crude oil suggest that the performance of these antifoams with pipeline crude oil (relative to that of unsubstituted PDMS antifoam) will be better than suggested by experiments with stabilised crude oil alone. This is because pipeline crude oil contains light hydrocarbon not present in stabilised crude oil.

4. ACKNOWLEDGEMENT

The authors wish to thank the British Petroleum Co plc for permission to publish this paper.

REFERENCES:

1) I.C.Callaghan, C.M.Gould, A.J.Reid and D.H.Seaton, European Petroleum Conference (London) Proceedings, 1982, 155.

2) M. A. Camp and F.T.Lawrence, "Apparatus Measuring Foam Stability", British Patent Application No. 2158574.

3) I.C.Callaghan, F.T.Lawrence and P.M.Melton, Colloid and Polymer Science, in press.

4) I.C.Callaghan and E.L.Neustadter, Chem.Ind. 1981, 53.

5) A.F.Sharovarnikov, V.N.Tsap, A.Ya.Korolchenko and A.V.Ivanov, Kolloidnyi Zhurnal, 1982, 43, 808.

6) A.Monsalvé and R.S. Schechter, J.Coll.Int.Sci.,1984, 97, 327.

7) N.W.Alcock, D.J.Benton and P.Moore, JCS Far.Trans. I, 1970,66, 2210.

8) E.V.Laere, J.Oil and Colour Chemists Assn., 1984, 67, 299.

9) L.J.Bowman, Spec. Chem. 1982, 2, 4.

Anionic Surfactants

By N. E. Jones and J. W. S. Patterson*
SHELL CHEMICALS UK LIMITED, I NORTHUMBERLAND AVENUE, LONDON WC2N 5LA, UK

Our brief is to attempt to give an overview of anionic surfactants for industrial uses, a brief which we accepted on the one hand with pleasure since the sheer diversity of uses presents endless opportunities for a raw material supplier, but, on the other hand, with some diffidence since we recognise that each industrial application develops its own in-depth expertise which can readily challenge the generalities and simplifications presented in an overview.

SCALE

Let us look firstly at the size of activity we are involved with here. Firm statistics are difficult to come by, due to both the diversity and confidentiality aspects of the industry, but at least as approximations the following figures may be reasonable guides to West European consumption:–

Table 1 SURFACTANT CONSUMPTION IN WEST EUROPE

Total Surfactant use:	1 250 000 tonnes/year
for Domestic products:	850 000
for Industrial uses:	400 000
Anionic surfactants in industrial uses:	250 000

What we see then is that industrial uses consume perhaps 1/3 of surfactant production and that anionic types are the larger part with alkyl benzene sulphonates playing, as indeed they do for domestic products, a major role. In any event, industrial uses of anionic surfactants comprise a large business activity, not only for surfactant producers but also for those manufacturers whose products or processes depend on achieving appropriate surface active effects.

DIVERSITY OF EFFECTS

The effects we are trying to achieve do not of themselves look very formidable and could be generalised as:–

Table 2 Surfactant Effects in Industrial Applications

Surfactant Effects
Wetting/Spreading
Emulsifying
De-emulsifying
Foaming
De-foaming
Solubilising
Dispersing
Cleaning

The difficulties and complexities for the surfactant formulator arise due to the wide range of industries he tries to service, each having its own need of one or more of these effects in greatly differing circumstances of gas/liquid, liquid/liquid, or solid/liquid surface or interface. A list of some of the processes and products involved give a feel for the complexity.

Table 3 Surfactant Applications in Industry

Metal processing oils

lub oil additives

gasoline, diesel and jet fuel additives

corrosion inhibitors

grease thickening

textile wetting agents/processing aids/printing

pigment dispersion

mineral flotation

dust suppression

pulp/paper processing

leather tanning and fat liquoring

emulsion polymerisation of latex

agricultural emulsifiers/dispersants

oil/wax/solvent emulsifiers

foaming agents in concrete, gypsum, latex

fire fighting foams

oil field drilling operations

oil field/well head uses

enhanced oil recovery

scale inhibitors

metal degreasing

acid pickling

antistatic effects

dry cleaning

grinding aids

plasticising/fluidising of cement.

Against this background of such diversity of needs and applications it is not surprising to find that surfactant formulators have responded to the problems by using an almost equally diverse range of molecules. Now, bearing in mind that we are discussing here only the anionic class of surfactants, let us take a look at just some of the molecular structures involved.

RANGE AND HISTORY OF MOLECULES

From the earliest days we have had soap with its linear alkyl chain and carboxylate head group derived by saponification of the triglycerides in tallow fat and coconut and palm oils.

FIGURE 1

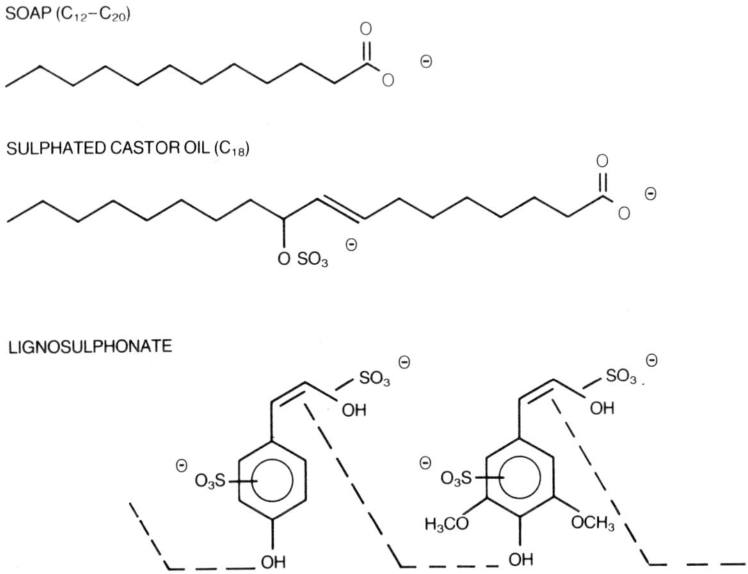

A variation of the anionic head group perhaps first appeared with the use of sulphated castor oil, the so-called Turkey Red oil, which was and may still be used in the formation of alizarin "lakes" for dyeing cotton. Here we see the introduction of a sulphate anion albeit in conjunction with a carboxylate. The alkyl chain is still linear.

However, looking at another early anionic molecule derived from naturally occurring feedstock, we see that neither carboxylation nor linearity have in any sense been limiting features of anionics. Lignosulphonates derived by sulphonation of tree rosin are used in great quantity as a source of cheap and presumably effective surfactant. The structure suggested in the Figure is grossly over-simplified but is intended to imply that aromaticity and branching of the hydrophobe and that the use of sulphate/sulphonate head groups offer a glimpse of the diversity of molecular variations that can be offered under the classification of anionic surfactants.

The last 50 years has of course seen more industrial development than all of the preceeding centuries added together. What has it brought in terms of molecules to the surfactant industry? Primary alkyl sulphates were recognised in the 1930's as alternatives to soap from the same natural fat feedstocks, and secondary alkyl sulphates were probably the first purposely developed products to be commercialised from the rapidly growing petroleum sourced feedstocks. Production started in 1942 and the molecule was derived from sulphation of olefins (chain length C_8–C_{18}) made by the thermal cracking of wax (C_{24}–C_{40}) which was a side product of refinery operations at the time.

FIGURE 2

PRIMARY ALKYL SULPHATE (C_{12}–C_{18})

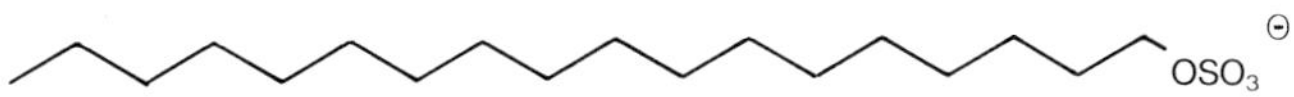

SECONDARY ALKYL SULPHATE (C_8–C_{18})

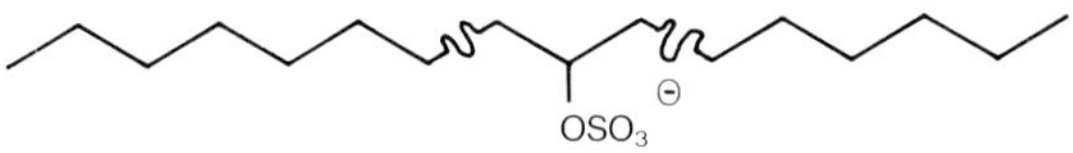

BRANCHED ALKYLBENZENE SULPHONATE (C_{10}–C_{14})

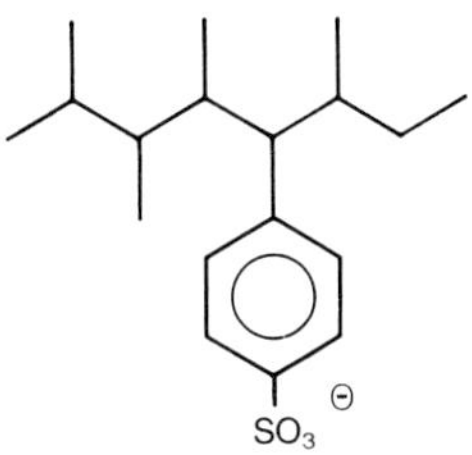

LINEAR ALKYLBENZENE SULPHONATE (C_{10}–C_{14})

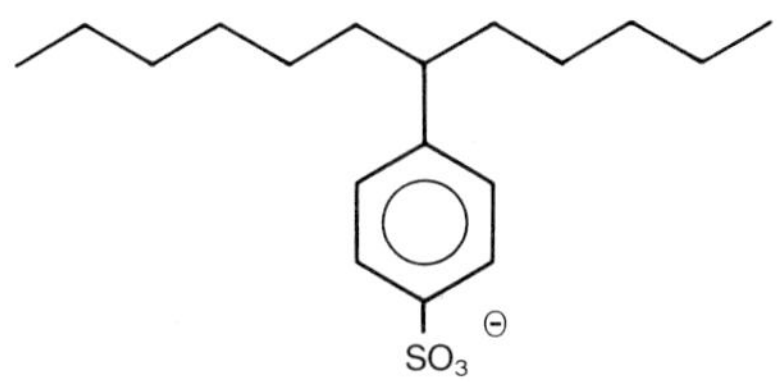

Post war developments enabled the rapid commercialisation of alkylbenzene sulphonate, based initially on the highly branched alkyl chain of propene tetramer which had been the source of war-time aircraft fuel, and subsequently on linear alkyl chains of approximately similar carbon number after it was realised that the replacement of soap as the major domestic surfactant lead to marked environmental repercussions as well as to the immediate goals of technical advantage and ready availability.

Also arising from the petroleum industry are several products which are widely used, but difficult to define chemically. They include petroleum sulphonates, naphthenates and alkyl naphthalene sulphonates.

FIGURE 3

PETROLEUM SULPHONATE

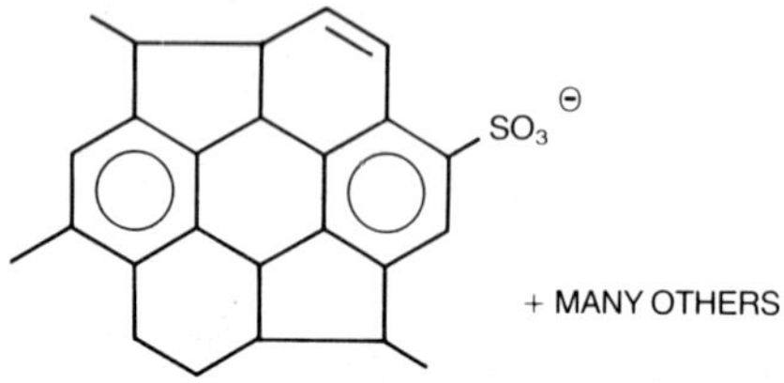

NAPHTHENATE

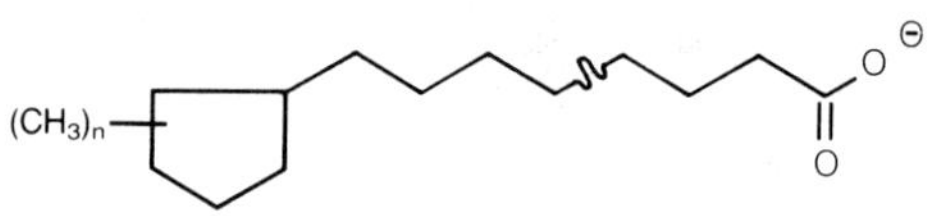

ALKYLNAPHTHALENE SULPHONATE

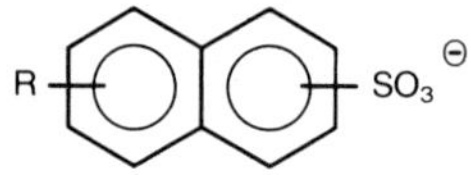

A fairly large number of anionic molecules of more specific chemical nature have been developed and brought into commercial use over the last decade or two. They derive from both petroleum industry sourcing and from natural fat feedstocks. Let us look at some of them.

FIGURE 4

SULPHOSUCCINATE

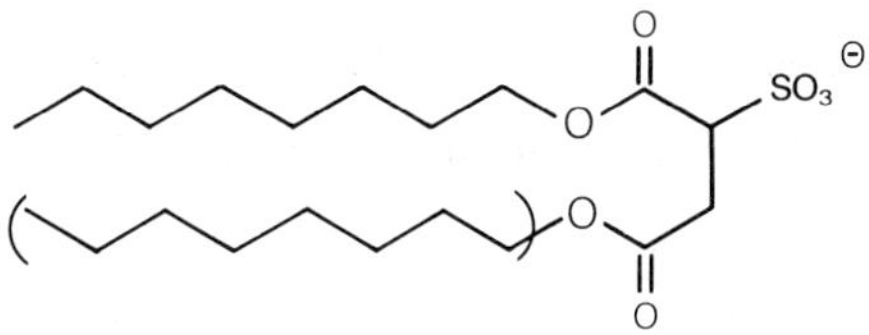

PHOSPHATE ESTER

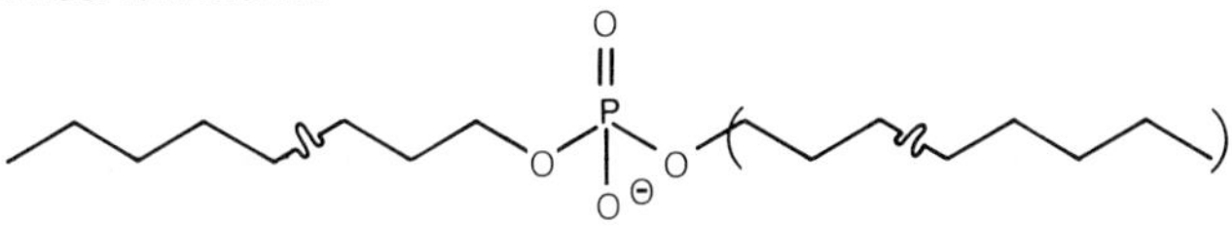

SALICYLATE

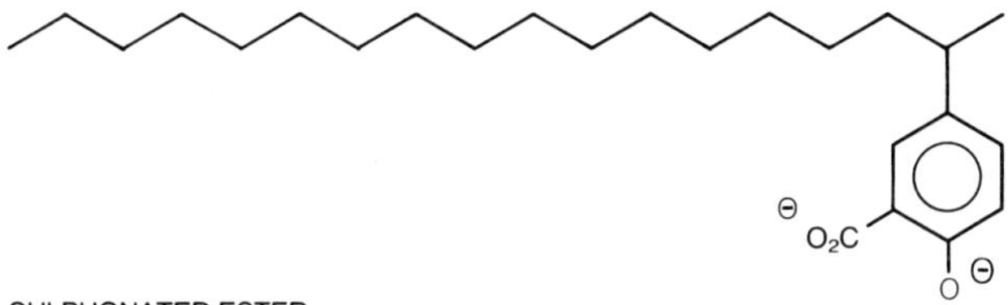

SULPHONATED ESTER

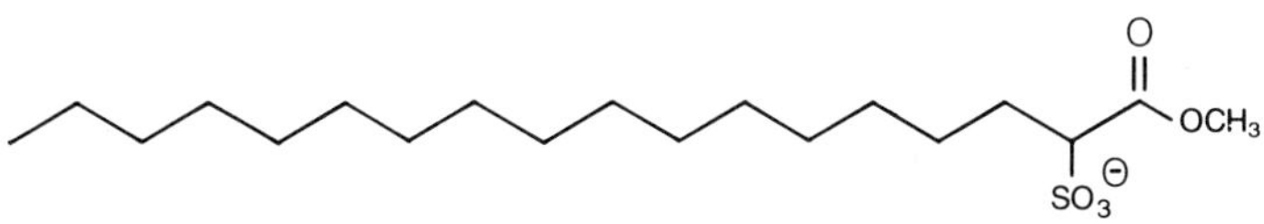

FIGURE 5

PARAFFIN SULPHONATE

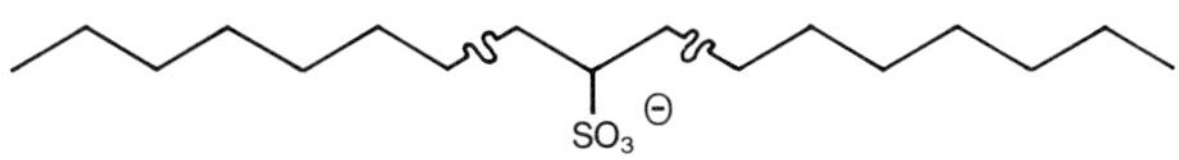

ALPHA OLEFIN SULPHONATE

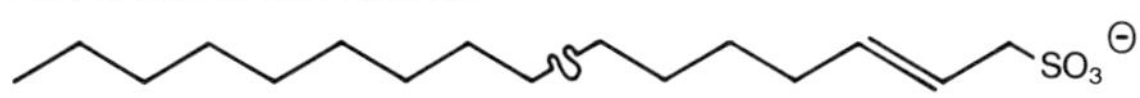

INTERNAL OLEFIN SULPHONATE

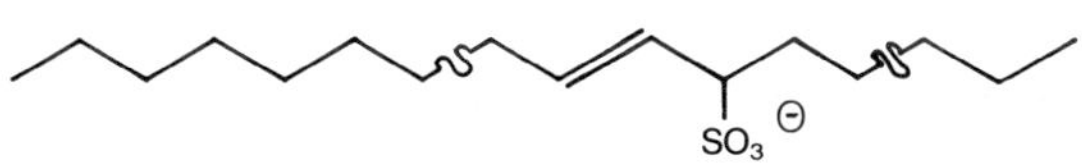

VINYLIDENE OLEFIN SULPHONATE

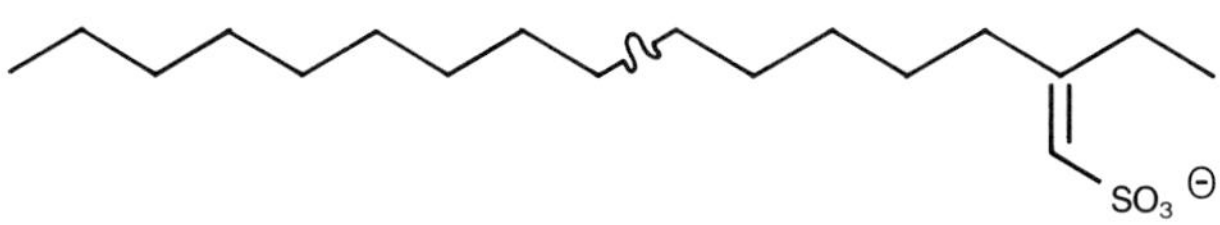

They include mono and di-alkyl sulphosuccinates, secondary alkane sulphonates (made by sulphoxidation of n–paraffins and having a C–S bond rather than the C–O–S bond of secondary alkyl sulphates made from sulphation of olefins) and various forms of sulphonate made from alpha, internal or vinylidene olefins and essentially leaving the double bond intact by virtue of substitution reaction with gaseous sulphur trioxide rather than addition across the bond, as would occur with sulphuric acid as reagent.

In rounding off this brief survey of generic structures, we must trespass a little into the territory of nonionic molecules since the addition of ethene oxide or propene oxide to a reactive site such as the hydroxyl group of an alcohol leads to a further site for sulphation which can markedly enhance the hydrophilicity of the molecule. Commercial examples of this approach include alcohol and alkylphenol ether sulphates, also referred to as ethoxysulphates or sulphated polyethers, and ethoxycarboxylates.

FIGURE 6

ALCOHOL ETHER SULPHATE

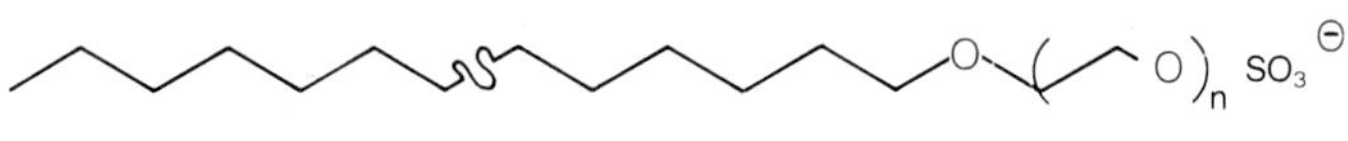

ALKYLPHENOL ETHER SULPHATE

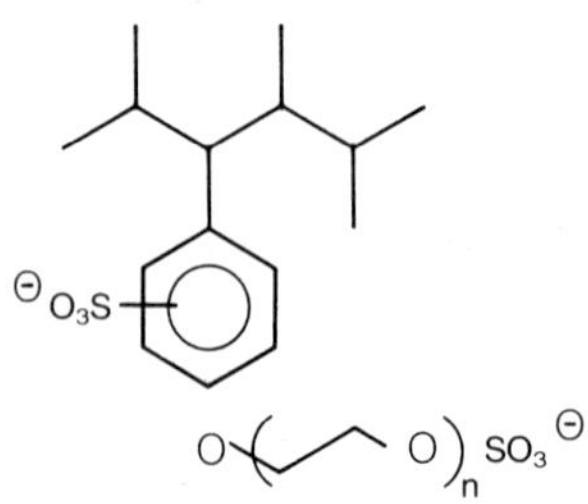

ETHOXYCARBOXYLATE

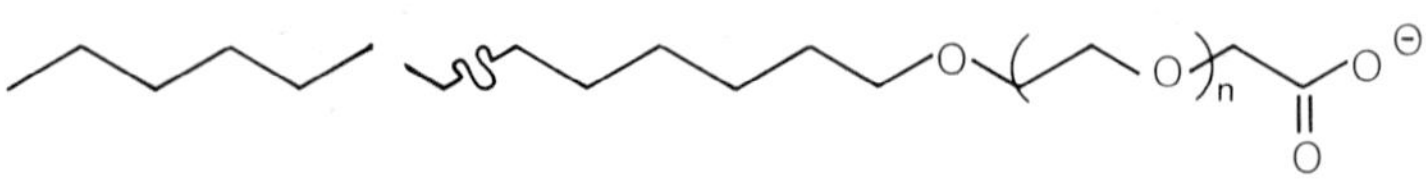

For reasons essentially based on ready commercial availability of feedstocks, the alcohol based products are mostly of linear alkyl form while the alkyphenol types are mostly of branched alkyl chain derived from either propene trimer (that is, C_9) or di-isobutene (that is, C_8). Some sulphonation of the aromatic nucleus is almost certain to occur whilst sulphating the ether group of alkylphenol ethoxylate.

If we stand back for a moment to reflect upon the situation, we can perhaps reach, or anyway jump-to, some conclusions. We have considered already 21 differing molecular species which could appear under the general heading of "Anionic Surfactants". I have no doubt that there are many more whose manufacturers and devotees will be offended by their absence from my summary. Our first conclusion must be that, if nothing else, anionics provide diversity; diversity of structure and diversity of application. I think a second conclusion is also self-evident. It is that such a wide range of products could only have been developed in response to a need. There has been a need in the solving of one or another industrial application problem to provide linear alkyl chains, branched alkyl chains, aromatic nucleii, different head groups, differing cations, greater or lesser hydrophilicity or hydrophobicity. Surfactant formulators have responded by reaching out for further variations of structure which would satisfy and overcome the problems presented to them.

IMPORTANCE IN APPLICATION

However, I want to come now to a third and more tentative conclusion which is rather more contentious. It is simply this; that most of our developments of industrial applications of anionic surfactants so far seem to have been based on an empirical assessment of their effects rather than a deep comprehension of the behavioural influences which would lead us to tailor a molecule to meet a specific circumstance. Thus, we have tried this and that and another molecule until we found one that achieved the effect we were seeking. There are justifiable reasons for such an approach. In the first place, one has to be assured of an economic and on-going feedstock and processing position for the surfactant since, although it is likely that the surfactant contributes but a minor proportion of the costs of a particular industrial application, failure to supply the right quantity of the right quality at the right time could cause substantial consequential losses due to stoppage of the process. To illustrate that point with one small example: gypsum based wallboard, or plasterboard as we call it in the UK, is made with a high volume of small diameter entrained foam bubbles which are stirred into the gypsum slurry before it is laid between the covering paper board sheets to set.

FIGURE 7

WALLBOARD MANUFACTURE

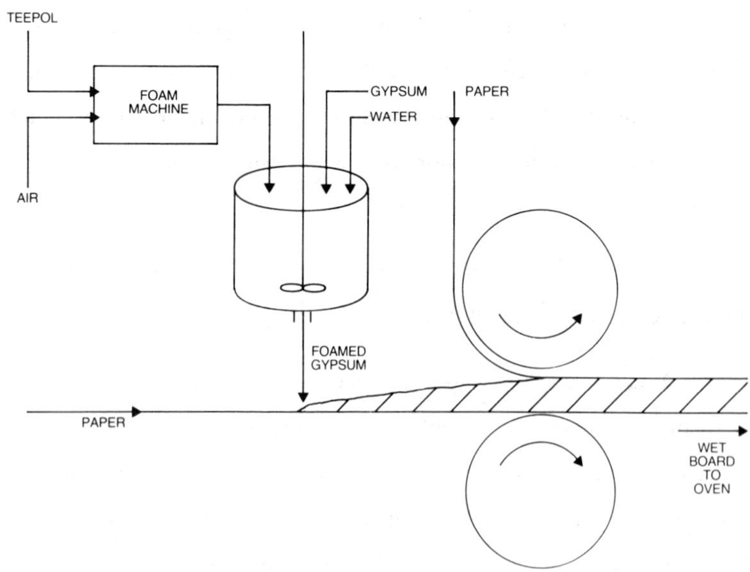

The amount of surfactant used to generate the foam does not reach percentage levels in the final board but without it the board would be very dense and wet, requiring a marked increase in drying energy costs, the gypsum would harden too rapidly causing cracking, and the composite would lack the thermal and noise insulation properties for which it is rightly valued. No doubt there are a myriad other examples which can be quoted with little thought.

In the second place, it is almost certainly true that domestic applications such as clothes washing and dishwashing have been given a higher priority in terms of surfactant development if only because they are large offtakes and were there first. These applications attracted, and no doubt still attract, a correspondingly large segment of the raw material and intermediate suppliers' R and D resources, together with capital investment programmes to make the required molecules. The industrial formulator has, not surprisingly, latched on to a number of these products to solve his own problems since they were readily available and at sensible cost. However, we should remember that the products were developed to meet the requirements of a fairly specific circumstance, that is, the removal of sebum and particulate soil from a fabric substrate in an almost totally aqueous environment. Whilst using them with gratitude we must also wonder whether they present the preferred molecules for many industrial uses with very different circumstances and environments.

POTENTIAL FOR TAILORING

I want to suggest that the major spin-off from domestic surfactant development into industrial applications has not yet been fully realised and grasped. It is not that there has become available a range of surfactant molecules in large quantity at near commodity prices. It is that in achieving that objective there has become available a complete infrastructure of technology, hardware, and supportive know-how which can offer the capability of tailoring molecules to meet specific performance criteria for industrial applications and in commercially viable quantities at an acceptable cost.

Let me try to illustrate that point. We tend to refer to alkylbenzene sulphonate as though it were a unique chemical with an unequivocal structure. The fact is that the product used, and I think very successfully, by the domestic detergent manufacturers is a blend of a number of related components. The alkyl chain is a mixture of chain lengths, generally in the C_{10}–C_{14} range, and for each alkyl chain there is a mixture of isomeric positions of attachment of the phenyl group.

FIGURE 8

'DOMESTIC' ALKYLBENZENE SULPHONATE

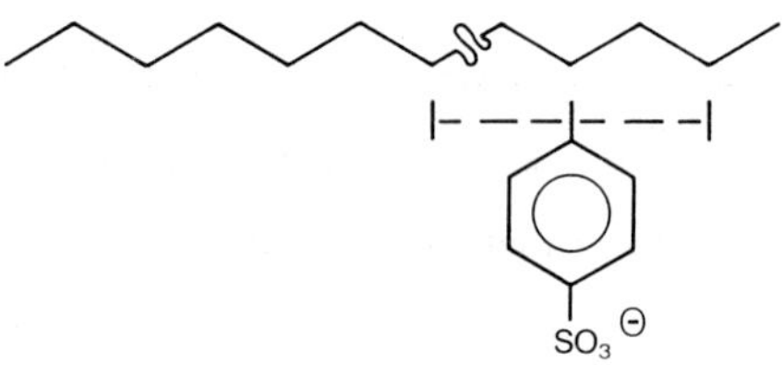

2–DODECYLBENZENE SULPHONATE

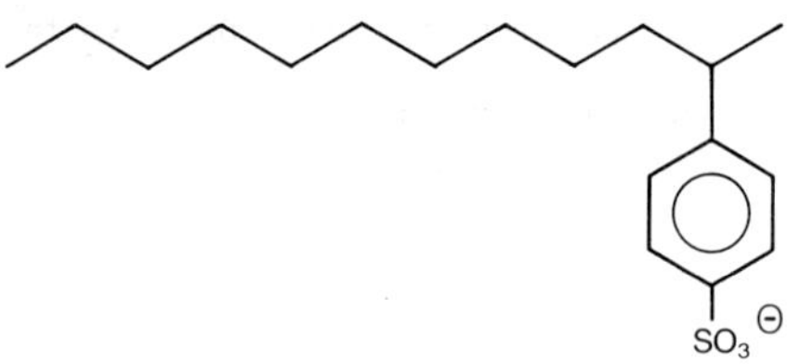

D I–DODECYLBENZENE SULPHONATE

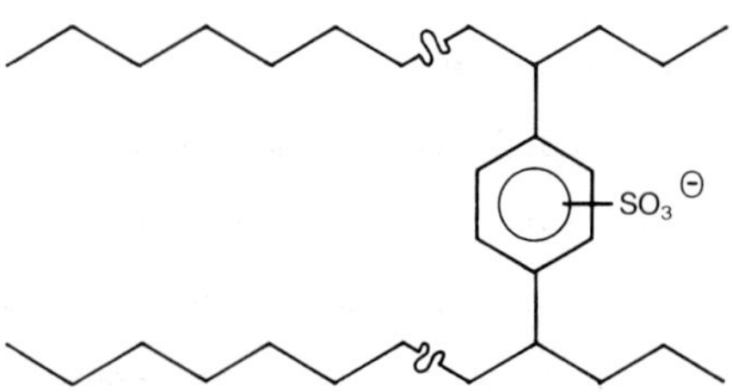

So whilst they can all be identified, the product has not less than 20 major components, in essence deliberately chosen for domestic cleaning. The development spin-off is that over the years the raw material suppliers have devised technology leading to narrow cut or even single carbon number alkyl chains and also to controlling to a large extent the positional attachment of the phenyl group. As far as I know, no commercial product provides l-phenyl isomer sometimes shown in textbooks but we can load the addition heavily onto the number 2 carbon of the chain. Interestingly, if we start with pure C_{12} alpha olefin and direct the benzene to the second carbon, the derived sulphonate has quite different water solubility characteristics to the broader cut and randomly isomerised alkylbenzene of the same average molecular mass. It is more hydrophobic.

Following the theme through, it is similarly feasible to re-alkylate the initial alkylbenzene to create the di-alkyl benzene. We then have a product of much greater molecular mass which has high solubility in oil, whilst retaining the polarity of the sulphonate anion. The product is starting to behave more like petroleum sulphonate than like domestic alkylbenzene sulphonate. Compare with Figure 3. The difference is that it has a highly defined and consistent composition which can be appropriately tailored to meet specific needs.

HARNESSING THE RESOURCES

Let me complete this illustration by referring to a fairly new range of alkyl aromatics which have arisen from the developmental theme I have just outlined. Bearing in mind that the aromatic base could be varied to be toluene or xylene over and above any variation in alkyl chain length and positioning on the ring, Shell Chemicals have adapted their proprietary alkylation technology to offer a series of homologous alkylxylenes and the sulphonated derivatives.

FIGURE 9

'DOBONAX' SULPHONATE (ALKYL O-XYLENE)

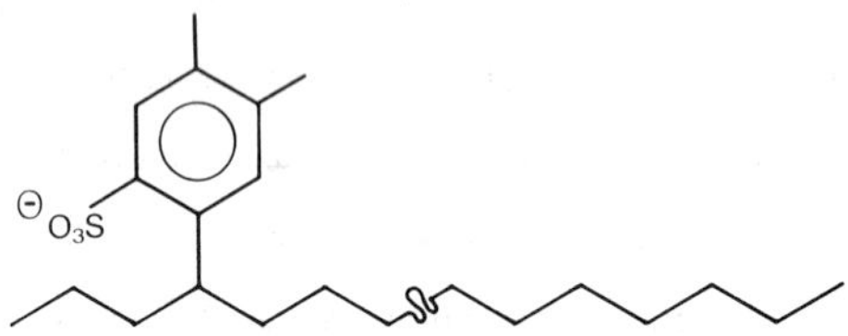

The products are based on o-xylene and SHOP linear olefins and the molecular structure is such that high oil solubility is achieved even with the lower molecular mass grades. The main isomer structure places the sulphonate moiety and the long chain alkyl group in the ortho position which directly influences the packing of the surfactant molecules at an oil/water interface and should be compared with the dominant para position placing of the sulphonate moiety of classical alkylbenzene sulphonate. The advantages over conventional petroleum sulphonate are much lower viscosity leading to ease of handling, the availability of the acid forms so that options can be exercised on the counter ion, and the choice of alkyl chain length. The range includes alkylates, sulphonic acids, and sodium sulphonates in oil and is offered under the trademark "Dobanax". Applications in hand include metal cutting oils, luboil additives, and oilfield operations.

There is no doubt in my mind that this molecular tailoring approach offers countless other opportunities across the field of industrial applications of surfactants and based essentially on the two parameters of (i) understanding what molecules are required (ii) adaptation of large scale technology to provide them economically. Such molecules are likely to carry some form of premium over domestic commodity surfactants but should prove significantly more cost-effective in their designated application.

Surfactants and Emulsion Polymerisation

By R. Groves
DOVERSTRAND LIMITED, TEMPLE FIELDS, HARLOW, ESSEX CM20 2BH, UK

Summary

Emulsion Polymerisation is an important technique for the production of polymers on an industrial scale. The product is a fine dispersion of polymer in an aqueous medium - a latex. Surfactant usually plays an important part in the process, particularly in the formation of particles and the maintenance of a stable dispersion during reaction.

Several factors need to be borne in mind when choosing a surfactant for an industrial latex system. The nature of the surface active species, whether cationic, anionic or non-ionic, is clearly important since this will almost certainly determine the nature of the latex produced. Surfactant Critical Micelle Concentration can be a useful guideline when designing a system to give a particular latex particle size. After these initial considerations, latex samples are produced to test the performance of various surfactants experimentally. Properties which may then be used to assess performance would include the latex stability in reaction and in application, its wetting ability and foaming tendency. The impact of the surfactant on application properties at the final point of use of the latex and its effect on latex cost would also be important factors in the considerations. The aim of the latex manufacturer is to identify a surfactant which will give trouble free production together with satisfactory final latex colloidal and polymer properties at minimum cost.

INTRODUCTION

During the last forty years, polymers in the form of stable, water based dispersions-latices - have become industrially extremely important. Applications for these products include paints, tyres,

paper coatings, textile binders, adhesives and plastics. The aim of this paper is to give a brief description of the emulsion polymerisation system by which these latices are made.
Some factors influencing the choice of surfactant in an emulsion polymer will then be discussed. This paper will be restricted to polymer dispersions in aqueous media, which are the systems presently of most industrial significance.

EMULSION POLYMERISATION

Free radical addition polymerisation is almost always the mechanism used in aqueous emulsion systems.[(1)] The polymerisation process is the linking together of a large number of small molecules (monomers) to form a large molecule (polymer). The main steps in the process are initiation, propagation, transfer and termination as illustrated in Figure 1.

In industrial emulsion polymerisation the reactions are accomplished by stirring monomer possessing limited water solubility in an aqueous medium, normally containing some surfactant. After removing most of the oxygen (which inhibits the polymerisation) from the reaction vessel, a source of free radicals is added and the reaction proceeds. The source of free radicals is almost always of the chemical type, for example peroxy compounds.

The role of the surfactant during reaction has been a source of considerable discussion over the years. Without doubt, it contributes to the emulsification of the monomer in water before initiation of the reaction. It also contributes to the stability of the final polymer dispersion. However, its function in the polymerisation reaction, particularly in the formation of latex particles, has been the subject of debate. The simplest explanation is that polymerisation occurs in monomer droplets and the function of the surfactant in particle formation is just one of producing a monomer emulsion of a given droplet size. This cannot be the case, however, because monomer droplets typically have diameters of 1000 - 10,000 nm which is much larger than the 100 - 500 nm diameter range found for latex particles.

The classic theory [(2)-(5)] describing the reaction mechanism requires surfactant to be present in sufficient quantity to form micelles, which are aggregates of perhaps 50 to 100 surfactant molecules, assembled so the hydrophobic portions are on the interior

Figure 1

The main steps in free radical addition polymerisation

I• Initiator radical	+ M monomer	→	IM•	Initiation
IM• + M		→	IMM•	Propagation
IMM• + M		→	IMMM•	
etc.		→	〰M• growing polymer chain	
〰M• +	X – Y chain transfer agent	→	〰MX + Y• dead polymer product	Transfer
〰M• +	Z• any free radical species	→	〰MZ dead polymer product	Termination

and the hydrophilic portions face the aqueous phase. The micelles are capable of solubilising some monomer in their interiors and this is the case at the start of the reaction. In addition, a small quantity of monomer is molecularly dissolved in the water phase. However, the bulk of the monomer, perhaps 90-98%, is present in the form of monomer droplets formed by mechanical agitation and stabilised by surfactant. This equilibrium is shown schematically in Figure 2a.

When a source of free radicals is added to this system, an individual radical is much more likely to encounter a micelle than a monomer droplet since the total micelle surface area is much greater than the total droplet surface area. Reaction therefore proceeds within activated micelles, monomer being supplied to them by diffusion from droplets or unactivated micelles. The activated micelles grow to become polymer particles with surfactant adsorbed on their surface (Fig 2b). The particles are kept separate (stabilised) by electrostatic repulsion in the case of ionic surfactants and by steric effects in the case of non-ionic surfactants. As the polymer particles grow, the monomer

Figure 2 a

Initial equilibrium at the start of Emulsion Polymerisation using an anionic surfactant

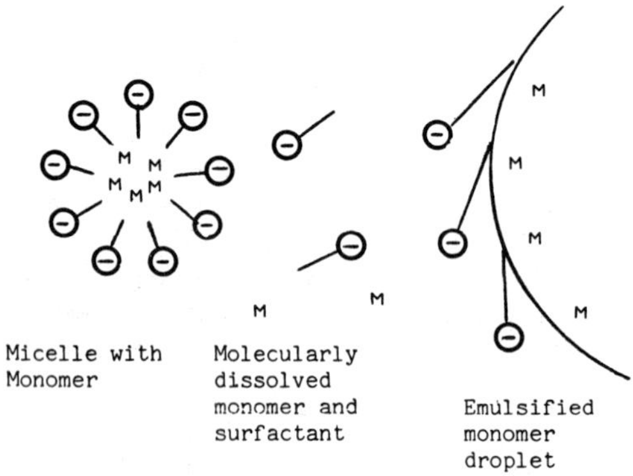

Figure 2 b

Final anionic latex

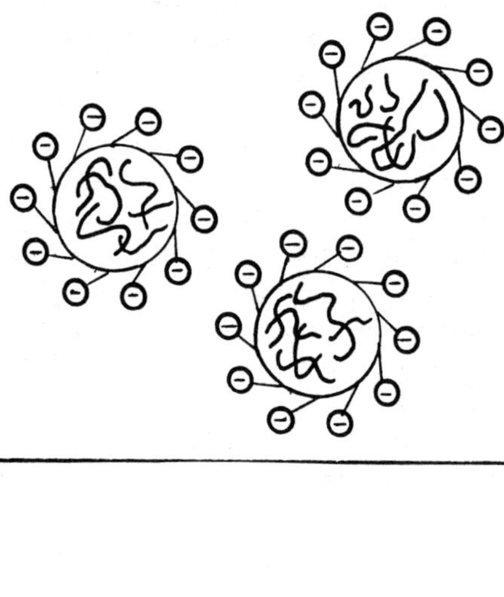

droplets and unactivated micelles lose their monomer and eventually disappear. This theory has been extended into quantitative treatments.[(6)-(8)] The details of these are beyond the scope of this paper, but one conclusion is that the number of particles formed per unit volume should be proportional to surfactant concentration to the 3/5th power.

Further experimental observations however, have led to the questioning of some aspects of this theory.[(9)] The 3/5 power relationship between particle number and surfactant concentration does not hold true for all monomer - surfactant combinations. In addition it is known that polymerisation can be achieved in aqueous systems which do not contain surfactant micelles, that is where the surfactant level is below its critical micelle concentration (CMC) or even in the absence of added surfactant. It has been proposed[(10), (11)] that under these conditions, particles are formed by the precipitation of oligomers (very short chain length polymer formed by polymerisation of the monomer in solution in the aqueous phase) to form primary particles, followed by coalescence of the primary particles. (Figure 3)

Figure 3

Homogeneous particle nucleation in the aqueous phase using anionic surfactant

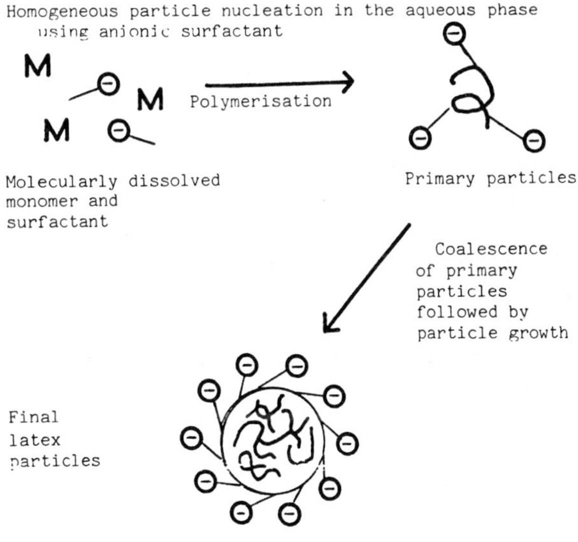

The primary role of the surfactant is envisaged as determining the extent of coalescence - increased surfactant would result in less coalescence before stability is achieved and thus an increase in the number of polymer particles per unit volume and a finer particle size latex. This mechanism might be expected to become more important with monomers of relatively high water solubility, for example vinyl acetate & methyl acrylate.[12] More recently it has been suggested that the coalescence mechanism for particle formation can be in operation even above the surfactant CMC and with less water soluble monomers such as styrene.[9] Quantitative treatment of the precipitation/coalescence theory is likely to be difficult, however.

To summarise, we have seen that the surfactant has a dual role in emulsion polymerisation. Firstly, it helps to form the initial monomer emulsion and secondly it stabilises the growing and final polymer particles.

Emulsion Compared to Alternative Methods of Polymerisation

It should be remembered that the Emulsion technique is only one of several industrially important methods of achieving the Polymerisation reaction. Other methods include Bulk, Solution and Suspension.

Many monomers can be polymerised in the absence of diluent, simply by the introduction of initiating species into the monomer. This is known as Bulk Polymerisation. It has the advantage of producing relatively pure polymer product directly. However, high viscosities are often encountered in the later stages of reaction which can lead to difficulty in removing the heat of polymerisation. This problem is particularly significant at large scale, making reaction control difficult. The molecular weight of polymer formed is inversely proportional to the rate of polymerisation, so the production of high MW material is difficult by this process.

Solution Polymerisation reduces the viscosity problem to some extent by using a suitable solvent for the monomer. However, if the polymer product is also soluble in the solvent, high viscosities and attendant difficulties can still be encountered. In addition, the solvents employed are generally expensive and difficult to handle from a health and safety point of view. As for the Bulk process, an inverse relationship exists between molecular weight and rate.

Suspension Polymerisation is essentially a compartmentalised Bulk Polymerisation, in which monomer is dispersed in a suitable medium (usually water) using polymeric stabilisers such as cellulose derivatives. A monomer soluble initiator is employed. Reaction takes place in the monomer droplets, in contrast to Emulsion Polymerisation, to give a dispersion with a particle size of 10,000 nm and above. This technique utilises a low viscosity reaction medium and heat dissipation problems are therefore overcome. Without agitation the polymer and dispersion medium will quickly separate assuming their densities are unequal. This can be an advantage if polymer is to be recovered close to the point of polymerisation, but a disadvantage if the suspension needs to be transported. As for the Bulk and Solution processes, the production of high molecular weight suspension polymer is difficult.

It will be recalled that Emulsion Polymerisation also uses a dispersion technique to give improved heat transfer. It is characterised, however, by the formation of high molecular weight polymer in a rapid but controlled reaction. The reason for this phenomenon is the subdivision of the reaction into many sites each of a small size - consideration of the kinetics of this system[6,1] leads to the conclusion that both polymerisation rate and number average degree of polymerisation are proportional to the number of particles. Therefore an increased number of particles gives a simultaneous increase in reaction rate and molecular weight.

Latices made by Emulsion Polymerisation have a small particle size, typically 100-500 nm, which is small enough for them to remain in suspension even without agitation and where there is a difference in density between the polymer and the dispersion medium.

Emulsion Polymerisation therefore produces rapidly a high molecular weight polymer in a low viscosity aqueous medium. It is this combination of properties which has resulted in the wide industrial use of latices. Two main drawbacks of the system should be mentioned, however. Firstly the presence of significant quantities of ancillary ingredients, particularly surfactant, can adversely affect the polymer performance in its final point of use. Secondly, a latex normally contains about 50% water which adds to product transport cost and usually requires the expense of drying equipment for its removal in the final application.

THE CHOICE OF SURFACTANT IN INDUSTRIAL EMULSION POLYMERISATION

The range of commercially available surfactants is wide and it is difficult to be sure that a given surfactant is indeed the best for a particular system. However, certain properties are usually required from a latex product so in practice surfactant choice can be based on the successful attainment of these properties.

The first decision to be made is whether an anionic, non-ionic or cationic hydrophilic group is required. This choice normally determines the character of the resulting latex. Most commercial latices are anionic, derived from the use of ingredients (including surfactant) containing anionic groups. It should also be mentioned that it is common industrial practice to use mixed surfactant systems, particularly non-ionic/anionic blends to achieve desirable stability properties. Non-ionic surfactants are useful in providing stability in the presence of cations such as Ca^{2+} or Al^{3+} which would destroy an anionic system, while anionic surfactants are effective in rendering a latex stable to mechanical shear.

The second general consideration made when selecting a surfactant is the nature and relative size of the hydrophilic and hydrophobic groups, together with the related solubility properties of the surfactant. By this means, the effective pH and temperature range of a surfactant can be estimated. Despite the more recent theories of mechanism, critical micelle concentration is often used as a guideline when choosing a surfactant to give a particular latex particle size.

In general C.M.C. decreases with an increase in chain length of the hydrophobic group for a particular surfactant type. An example is given in Figure 4 where the C.M.C. values for a series of carboxylic acid soaps are shown.[13]

For a given molar surfactant concentration, a decrease in C.M.C. would result in more micelles, and hence a larger number of smaller sized latex particles. This in turn leads to an increase in reaction rate because of the increased number of reaction sites (particles). Overall, a greater chain length in the hydrophobe should therefore result in an increased reaction rate. Figure 4 also shows reaction rates for a series of styrene-butadiene latices stabilised by the same fatty acid soaps, each used at a constant weight of 5% with respect to monomer.[14]

Figure 4

C.M.C. values for an homologous series of potassium fatty acid soaps [13] and initial (8 hr.) reaction rate found when they are used in a styrene-butadiene polymerisation [14].

Hydrophobe chain	CMC (mole/litre	(g/l)	T (^{o}C)	8 hr av. reaction rate at 50°C (% conversion per hr.)
C_{10}	0.10	21.0	25	1.8
C_{12}	0.025	6.0	20-25	3.4
C_{14}	0.0065	1.7	20-25	5.3
C_{16}	0.0017	0.5	35	6.4
C_{18}	0.00044	0.14	50-60	6.4

Over the range examined, the expected trends of initial reaction rate with hydrophobe chain length are seen.

The next step in surfactant selection is essentially a process of trial and error. Materials thought suitable from the initial considerations are used to produce experimental latices. Polymerisation characteristics and final properties of the latex are then checked against those required from the product. Given below are some examples of important factors by which the effectiveness of a surfactant might be judged.

a) Stability

Quite clearly a latex must be capable of being produced without giving stability problems. Insufficient stability in reaction manifests itself as the formation of lumps of polymer, known as grit or coagulum, or even the total solidification of the reactor contents. The correct choice of surfactant can make a significant contribution to reaction cleanliness. [15] To illustrate this, Figure 5 shows the effect on reaction cleanliness made by using various anionic surfactant types in a particular emulsion system. In this case a certain order of cleanliness is seen, but it should be stressed that this effect is dependent on the emulsion system under investigation; another system would probably give a different order of cleanliness.

Figure 5

The effect on coagulum level of changing surfactant type in one particular polymerisation system. Styrene-butadiene monomers, surfactant at constant level by weight.

Surfactant type	Coagulum level (% dry on dry polymer in latex form)
alkyl benzene sulphonate	1.7
sulphosuccinate	1.4
alcohol sulphate 1	1.1
alkylated oxy dibenzene disulphonate	0.8
alcohol sulphate 2	0.1

After reaction, additional surfactant may be added if needed to confer the correct stability to a latex. In some cases, for example paper coating or addition to cement, very high stability is required to withstand the hostile environment to which the latex is subjected. Non-ionic surfactants can be useful additives for these applications to assist in achieving high stability to multivalent cations. In other areas, for example tyre rubber production, minimal stability is needed since just after polymerisation the latex is deliberately coagulated to give a crumb. High stability in this case would be a disadvantage. Carboxylic acid soaps might be selected for this application, so that coagulation can be effected simply by reducing latex pH.

There are also some specialised applications where a latex is required to coagulate to give a solid gel when its temperature is raised to a particular level. The impregnation of non-woven fabrics with polymer in latex form is one example of such an application.[16] Temperature controlled coagulation of the latex prevents migration of the polymer particles towards the fabric surface on drying, resulting in a more uniform product. This property is obtained by the use of non-ionic surfactant of the ethylene oxide condensate type. Figure 6 gives results from a series of latices made with 3% of C_{16}-C_{18} alkyl alcohol - ethylene oxide condensate surfactants. Variation in the length of the E.O. chain indicates that in this case an approximately 10-15 mole type of product gives the desired heat sensitisation temperature of 50°C. Here again, the results obtained are highly dependent on the composition of the total system - electrolytes, other surfactants, sensitising agents can also have major effects.

Figure 6

Gelation temperature of latex containing 3% of C_{16}-C_{18} alcohol - EO condensate

No of moles EO in surfactant	Latex Gelation temp.(^{o}C)
50	> 70
20	58
15	50
11	47
10	49
6	> 70

From the above examples it can be seen that different applications make different stability demands on a latex. These range from the minimum requirement of being able to produce a clean product from a reactor up to the ability to withstand direct addition to dry cement.

b) Wetting

Latices are often used in coating operations. Some of the substrates which need to be coated are difficult to wet with water based systems. An extreme example occurs in the production of pressure-sensitive adhesive products by a transfer process. Here the adhesive, in latex form, is first spread onto a low surface energy release paper and after drying, the adhesive film is transferred to the intended substrate. Spreading onto the release paper to give a thin, coherent latex film can be difficult; a reduction in latex surface free energy by the addition of a suitable surfactant is desirable, together with latex viscosity control. It should be remembered of course, that surfactant additions can lead to latex foaming which may be undesirable in many applications.

c) Foaming

Most latices have a tendency to foam and this can cause difficulties for the latex manufacturer. The difficulties are normally overcome by the addition during processing of mineral oil derivatives which are available as antifoams. When a latex comes to be used in its final application, foam formation may need to be suppressed by further antifoam addition. However, the correct choice of surfactant and surfactant level can minimise the need for antifoams.

For some processes, aqueous latex based mixes are deliberately foamed by mechanical means prior to use. One example is the foam backing of carpets, where relatively high levels of surfactant are employed in order to achieve easy air incorporation and subsequent foam stability. Figure 7 illustrates the improvement in foamability which the addition of sodium lauryl sulphate makes to a carpet backing latex.

Figure 7

Foamability improvement by adding sodium lauryl sulphate to a carpet backing latex. Latex already contained 0.6 parts sulphosuccinate.

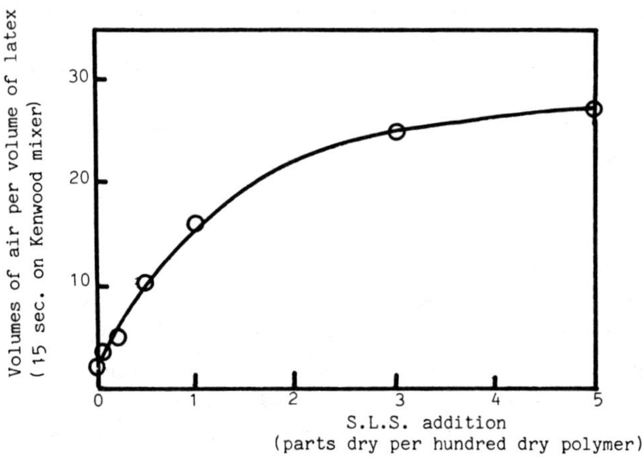

It can be seen that S.L.S. is efficient in promoting foamability, enabling over 25 volumes of air to be incorporated per volume of latex. For this reason it is often used as a foam booster in carpet backing latex formulations.

d) Particle Size/Viscosity

The level of a particular surfactant used in an emulsion polymerisation, as we have discussed, influences the particle size of the latex produced and thus the reaction rate. For all surfactants, increased amounts lead to reduced particle sizes, except where very high levels are employed. The results from an experimental series are shown in Figure 8, where the particle

Figure 8

Particle size dependence on surfactant concentration in an experimental styrene butadiene latex series.

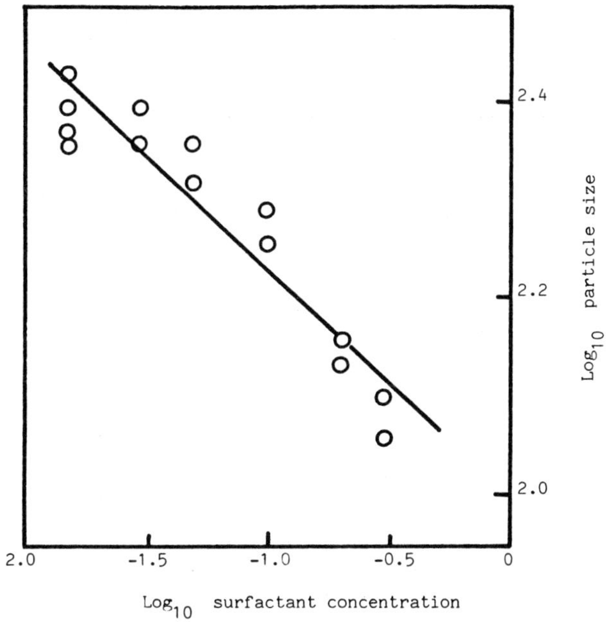

size dependence on surfactant concentration for a particular latex system is plotted on a log scale.

It is interesting to note that in this case, if the best fit straight line is imposed, then the slope corresponds with the 3/5th power relationship mentioned in the first section of this paper.

For a given weight fraction of polymer, the latex particle size plays an important part in determining latex viscosity (usually measured at low shear rates). Reduced particle size leads to increased viscosity as shown in Figure 9. In turn, high viscosity is normally the limiting factor in determining the maximum weight fraction at which it is practical to supply a latex commercially. Clearly an increased particle size, which might be achieved by a low surfactant level, would result ultimately in an increased solids content product with consequent advantages in transport costs and ease of drying at the final point of use.

Increasing a latex particle size can have disadvantages, including slower reaction rates for the manufacturer and, in some applications, poorer performance. In practice, therefore, a particle size is sought which gives the best compromise between the various properties required.

Figure 9

The effect of latex particle size on viscosity. Latex samples all 50% solids content.

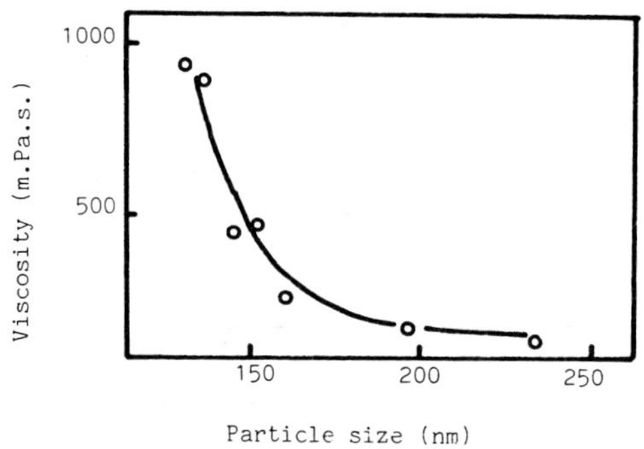

d) Properties of the Final Product

The applications for latices are widespread and so generalisations concerning the effect of surfactant on product properties cannot be made. However, during recent years it has become increasingly clear that in certain cases, surfactant used in emulsion polymer production can have a deleterious effect. Two examples may serve to illustrate this.

As mentioned earlier, many non woven fabrics are held together with a polymer which is introduced in latex form into the fabric. For some uses, water resistance of the final fabric is required. A simple test for this property is to place a drop of water on the fabric and measure the time taken for the drop to soak in. Figure 10 shows data from the same fabric impregnated with a standard latex compared to other latices made with reduced surfactant levels.

Figure 10

Time for a water drop to absorb into fabric impregnated with various latices

	Approx. surfactant level (%)	Absorption time (sec)
Standard latex	5.0	Instantaneous
Latex A	1.5	Instantaneous
Latex B	1.0	Instantaneous
Latex C	0.6	100
Latex D	0.5	850

It can be seen that with the higher levels, water repellancy is non existent and benefit is only obtained when significant reductions are made.

A clay coating is often applied to high quality paper to provide a smooth surface for printing. Most coatings contain polymer in the form of latex to bind the clay particles to the paper fibres. The coating needs to have a strength sufficient to withstand high speed printing with viscous inks. In some printing processes, the coating is wet as it is printed, so wet strength is also important. Wet strength values from a series of latices containing various levels of surfactant are shown in Figure 11; again it can be seen that results worsen with increased surfactant.

Figure 11

Effect of surfactant level on Wet Strength of a Paper Coating Latex

	* Wet Pick (% Ink Density)
Latex A	90
Latex A + 0.6% SLS	78
Latex A + 1.2% SLS	56
Latex A + 1.8% SLS	43

* Higher figures indicate better wet strength

The reason for the effect of some surfactants on properties is a matter for speculation. It has been shown that certain surfactants exude to the surface as the latex polymer particles coalesce.[17] In a non-woven fabric or paper coating this surfactant will then be available to promote wetting as water is applied. In addition, the exudation process might cause disruption to the polymer film or the polymer-substrate interface giving rise to weak points and an overall reduction in strength.

To try to overcome some of these problems, much work has been carried out by manufacturers of emulsion polymers aimed at producing latices which are stable and yet contain low emulsifier level. This effort has shown some positive results; a number of low emulsifier products are commercially available and are being used with success.

e) Cost

There is currently a greater pressure than ever on producers to keep the cost of emulsion polymer systems to a minimum. The prices of different surfactants vary widely and using one in preference to another may make a difference of up to 10% on latex cost. The economic consequences of using a particular surfactant, therefore, should be at the forefront of the mind of every industrial latex chemist.

CONCLUSION

It has been seen that surfactant plays an important part in emulsion polymerisation reactions, particularly in the particle formation stage. The type and level of surfactant is decided with reference to the important properties which a latex should possess, especially its particle size and stability. The product made from the latex can also be affected by the presence of surfactant and is an important consideration in development work.

In the future, it is likely that latex development will continue to seek improved stability at minimum cost, suggesting low levels of efficient surfactants. These may be made separately from the polymerisation, or in situ as the reaction proceeds. It is generally true that much industrial research time is spent on reducing coagulum or reactor build-up in otherwise good latex formulations. A better understanding of coagulum and reactor fouling mechanisms coupled with the development of surfactants which reduce these problems would find wide use in the emulsion polymer industry.

The author wishes to thank the Directors of Doverstrand for permission to publish this paper and Mr. G.C.N. Cheesman for help in its preparation. Most of the results quoted were generated by the various Technical Groups within Doverstrand whose work is gratefully acknowledged.

References

(1) D.C. Blackley "Emulsion Polymerisation" Applied Science 1975

(2) W.D. Harkins J. Chem. Phys. 1945 13 381

(3) W.D. Harkins J. Chem. Phys. 1946 14 47

(4) W.D. Harkins J. Am. Chem. Soc. 1947 69 1426

(5) W.D. Harkins J. Polym. Sci. 1950 5 217

(6) W.V. Smith & R.H. Ewart J. Chem. Phys. 1948 16 592

(7) J.L. Gardon Br. Polym. J. 1970 2 1

(8) J.L. Gardon Rubb. Chem. Technol. 1970 43 74

(9) A.S. Dunn Makromol Chem. Suppl. 1985 10/11 1

(10) C.P. Roe Ind. Eng. Chem. 1968 60 20

(11) R.M. Fitch & C.H. Tsai "Polymer Colloids" Plenum 1971

(12) J.W. Vanderhoff J.Polym. Sci. (Polymer Symposium) 1985 72 161

(13) J. Brandrup E.H. Immergut(Ed) Polymer Handbook (2nd Ed) John Wiley 1975 II 465

(14) C.W. Carr, I.M. Kolthoff, E.J. Meehan & D.E. Williams J. Polym. Sci. 1950 5 201

(15) J.W. Vanderhoff Sci. Technol. Polym. Colloids 1983 67 167

(16) T.J. Woods Paper 15 presented at "Polymer Latex" (PRI) London, 1st November 1978

(17) E.B. Bradford & J.W. Vanderhoff J. Macromol Chem. 1966 1 335

Interaction of Fire Fighting Foams with Burning Hydrocarbons

By A. A. Briggs
BUILDING RESEARCH ESTABLISHMENT, FIRE RESEARCH STATION, BOREHAMWOOD, HERTS. WD6 2BL, UK

As a fire extinguishing agent water has many advantages. The high latent heat of evaporation is a valuable asset while, in general, supplies are plentiful and inexpensive. On fires of liquid hydrocarbons, water spray can reduce flame intensity. With involatile fuels it may be possible to cool the fuel below its fire point. However, because water is denser than virtually all hydrocarbons, it cannot seal the surface of fuels. The use of foaming agents in water has the primary function of enabling a reduction of density. There are secondary benefits including a limited ability of foam to cling to vertical surfaces.

All modern fire-fighting foams (with one recent exception) are made by mechanical incorporation of air (or, for special purposes, inert gases) into foam agent solution. (Foam agents using chemical reaction to generate carbon dioxide may be regarded as obsolete.) In all that follows in this paper, 'foam' is used as an abbreviation for fire fighting foam while 'fuel' covers the hydrocarbons found in petrol and light oils eg kerosene and diesel fuel.

Classification of Foams

It is convenient to classify foams by two methods:

(a) by expansion ratio

(b) by chemical origin

Expansion ratio is defined as the ratio of the volume of foam to the volume of solution from which it was generated.

The categories, somewhat arbitrarily defined, are:

(1) low expansion : up to 20

(2) medium expansion : 20 - 200

(3) high expansion : > 200

Alternative figures will be found for this categorisation but the above have been recommended for international standards.

Chemical classification arises from the derivation of the foaming agent. The classical foam agents are derived by hydrolysis of proteinaceous vegetable and animal materials and are loosely known as 'protein foams'. More recently, detergent type foams were introduced (colloquially 'synthetic'). In the early 1960's fluorochemical surfactants were incorporated into these bases leading to new categories of 'fluoro-protein' and 'fluorochemical'. Commercial examples of this last type usually have a positive spreading coefficient (see later) for the aqueous solution over hydrocarbons leading to the expression 'aqueous film forming foam' (often abbreviated to AFFF).

Methods of Application of Foams

Most foam is used for fighting fires of hydrocarbon spillages. Within this category most of this foam will be applied from hand-held, vehicle mounted or fixed/oscillating 'branchpipes' or 'monitors' and will have low expansion ratio, typically 6-10. The agents used are likely to be either protein based or AFFF though, in a few countries, 'synthetic' agents are used in this form of application. The use of medium or high expansion foam in this type of fire situation is limited by the short projection range and vulnerability to wind including upsweeping air currents generated by fires.

Low expansion foams may also be applied through sprinklers, foam pourers mounted on storage tanks or by base injection into these tanks. This last method is a positive way of ensuring that foam is actually applied to a burning tank (as distinct from being carried away by wind or flame-induced up-currents). Interaction of foam and hydrocarbon is likely to be severe in this form of application.

Foam agents based on 'synthetics' tend to be more easily foamed than the protein-based agents, ie simpler equipment will give adequate foam for some purposes. However such foams also tend to lose water more readily and therefore show less thermal resistance. The need for the latter varies with the nature of the fire. This is just one controversial area of fire/equipment/agent interaction. There is no great difficulty or penalty in making an efficient foam-making branchpipe. Equally if simple equipment does suffice, there is unlikely to be any objection to its use.

High expansion foam, of expansion ratio of 500-1000, was first used in coal mines.[1] At first sight a water content of 0.1% v/v seems very low, but it does represent c 60% v/v when the water is evaporated, ie oxygen concentration in the atmosphere is reduced from 21% to 8%. This reduction is sufficient to suppress flaming combustion (subject to containment of this atmosphere) but not smouldering. Flooding or other water damage is thereby minimised[2]. High expansion foam filled with nitrogen has been used to provide protection against possible explosion where hot work is required on tanks containing flammable residues[3].

The role of medium expansion foam is currently ill-defined. It has been used with some success on fires of rubber tyres, subject to its limited projection range which is normally less than 10 m. Both high and medium expansion foams use 'synthetic' agents.

Since effective projection of any extinguishing agent to the seat of the fire often presents difficulty, there is a case for high density (very low expansion) foam. This may be extrapolated to a stream of agent which forms foam only after impact at its target. Recently a product ('Halofoam') has been demonstrated which shows certain abilities in this requirement. This agent is said to contain fluoro-chloroethanes suspended in a relatively viscous commercial foam concentrate.

Operational Aspects of Foam Application

It is thus established that most foam is generated in transportable nozzles which yield low expansion foam for projection on to hydrocarbon fires ranging from a few mm deep ('spill' fires) to 20-25 m deep ('tank' fires). In these nozzles foam agent is pumped through an orifice at high pressure. The resulting spray entrains ('aspirates') air as it passes along a tube 0.5-1.5 m long and emerges as foam with a velocity of 10-20 m/s. (For certain large apparatus higher figures may apply.) It may be of significance to foam generation and subsequent stability that the velocity of sound in typical low expansion foams will be only 30-40 m/s.

Where a blanket of foam can be applied gently and evenly to the fuel surface then interaction with fuel will be minimal. In practice this is a rare occurrence. Access to a fire may be poor, limited visibility may leave the situation ill-defined while, outside the specialist fire brigades, training and experience may be limited. In practice foam is likely to be applied with a considerable impact velocity on the fuel. In view of the complexity of analysing turbulence phenomena in a heterogeneous foam/fuel system, simpler parameters must suffice.

Physical parameters would include

1. Velocity of the foam stream or 'pellets' which in turn can be translated into
   (a) momentum/energy
   (b) local surface pressure

2. The deformability of foam which includes
   (a) viscosity (non-Newtonian)
   (b) coherence ie resistance to break-up on impact linked strongly to 1. (a)

The viscosity of the substrate fuel can be dismissed since with the volatile fuels under consideration the range of variation is small. Plates 1 and 2 illustrate entrainment of fuel by foam with a laboratory branchpipe operating at 6 m/s.

Chemical parameters are likely to include the foam agent composition and fuel composition. It is significant that with water, the interfacial surface tension of volatile aromatic hydrocarbons is significantly lower than for aliphatics. Present-day petrol contains up to 40% aromatics and this content may increase with the phasing-out of tetra-ethyl lead.

Formulation of Foam Agents

Since the mixing of fuel and foam must be regarded as inevitable in many fire situations, the ability of a foam to tolerate this fuel must be assessed. If we consider the spreading coefficients, defined by

Plate 1: Application of fluoroprotein foam to heptane

Plate 2: Application of AFFF to hephane

(a) Fuel on water $S_{f/w} = \gamma_a - (\gamma_f + \gamma_i)$

(b) Water on fuel $S_{w/f} = \gamma_f - (\gamma_a + \gamma_i)$

where $S_{f/w}$ = spreading coefficient of fuel on water
$S_{w/f}$ = spreading coefficient of water on fuel
γ_a = surface tension of aqueous phase
γ_f = surface tension of fuel phase
γ_i = interfacial tension

For both protein and synthetic foams, $S_{f/w}$ is positive ie fuel can spread on foam membranes. This situation favours both the physical spread of fuel and its evaporation. A specific illustration of this arises from the 1952 trials by the Fire Research Station of base injection of a protein foam[4]. Although fire intensity could be reduced the flammability of the foam blanket prevented complete extinction.

With increasing usage of liquid fuels, coupled with a specific disaster, the US Naval Research Laboratory investigated the scope for improvement of foams (ca 1960). New proposals took note of the characteristics of fluorocarbon chains ie that, in contrast to the hydrocarbon chains of most surfactants, the fluorocarbon chain was both hydrophobic and oleophobic. This was reflected in negative spreading coefficients for fuel on aqueous phase. The practical consequence can be seen in base-injection trials by Nash and Whittle [5]. In these trials a protein foam contaminated with 8.5% $v/_v$ (relative to foam solution) of fuel was readily ignited; the corresponding agent with fluorosurfactant addition, 'fluoro-protein', was not ignited in these conditions of test. Corresponding improvements in extinction performance by foam from branchpipes have been found.

With 'synthetic' foaming agents (and, more recently, protein-agents) fluorosurfactants can provide positive spreading coefficients for aqueous phase on fuel (AFFF). These aqueous films reduce the rate of evaporation of fuel and thus inhibit burning [6]. More recently these films have been found to accelerate ignition of less volatile fuels eg kerosene and diesel fuel[7]. This phenomenon is provisionally attributed to a 'wicking' effect.

While the formulation of foams is a topic of some interest to a chemist, the primary concern of the Fire Research Station is with the consequences of these formulations. Certain generalisations emerge from consideration of various sources. Thus much 'protein' foaming agent is currently produced by alkaline hydrolysis of hoof and horn meal. Addition of iron and/or zinc salts improves foamability and heat resistance. 'Synthetic' foam agents usually consist of anionics with solid alcohols, usually C_{12}, to stabilise the foam. A substantial list of fluorochemical surfactants can be gleaned from the patent literature [8] but, of those, one surmises that $n\text{-}C_7F_{15}COONa$ was one of the earlier components while some current formulations include $R_f\,CH_2CH_2SCH_2CH_2CONHC(CH_3)_2CH_2SO_3Na$, where $R_f = C_6F_{13}$ or C_8F_{17}. Given that foam agents incorporate less fluoro-surfactant than is required for, say, textile processing, the 'tailoring' of agents may be limited. Foam manufacturers must take account of cost, availability, stability in storage, toxicity/biodegradability corrosion characteristics etc. Some large contracts hinge on minimum requirements to meet a specification at minimum cost. Unfortunately the specification, while providing a useful assessment, does not guarantee an optimum combination of properties for full-size fires.

Evaluation of Foam Agents

Since 'good' foams may have widely differing and complex chemical compositions, chemical analysis cannot reliably define fire performance. Tests of the agent on fires become virtually inevitable.

For maximum reliability, the test should be performed on the hazard itself or a close simulation. This concept ranges from the expensive to the impracticable. If comparative tests are required, the variability of weather, notably wind and ambient temperature will lead to difficulties of comparison. As a consequence laboratory tests have been evolved.

Most of the relevant Standards require that foam is generated in a specified laboratory-scale branchpipe and its properties then assessed. As discussed earlier, the need for relatively efficient foaming by a branchpipe is questioned. The author takes the view that foam agents should be tested in an efficient branchpipe with a separate additional test for foam agents claiming suitability for 'non-aspirated' application.

The properties assessed are usually

(a) Expansion ratio

(b) Rate of drainage under specified conditions

(c) 'Viscosity'

(d) Performance on a specified fire

One well-known test requires[9], additionally, surface and interfacial tension measurements and even the conformity of refractive index of the solution.

The rate of drainage of aqueous solution from foam is obviously of significance as a factor relating to the thermal resistance of foam. Rates will depend on depth of foam but one widely used drainage vessel is 200 mm deep. In these conditions the foam can be expected to lose by drainage some 25% of its solution content in 4-8 minutes. AFFFs need to drain relatively rapidly in order to replenish the aqueous film.

Expansion ratio is, at first sight, a self-explanatory requirement. However foams with substandard expansion ratio and fast drainage tend also to 'disintegrate' when they are projected. Hence their range is limited.

Viscosity of foam has been discussed elsewhere at length[10]. The method of measurement provides one result in a series of complex inter-relationships. Arguably one result, obtained in a simple measurement, is better than none.

Fire performance tests and their derivation would need a paper to themselves. The situation may be summarised as follows:

(a) The standard methods often apply to one species of foam only and reflect pre-conceived ideas concerning its performance

(b) one common factor concerns the relatively low velocity of impact of foam on fuel. In tests familiar to FRS this is less than 5 m/s. Trained, if inexperienced, firemen have been known to apply foam with an impact velocity of about 18 m/s. The terminal vertical velocity of a 25 mm pellet of typical low expansion foam has been found to be 6 m/s

(c) in some standards, the choice of fuel (eg n-heptane) is probably unrealistic

(d) an international working group has decided that, rather than use any existing fire test methods, a fresh approach is required.

In a recent series of fire tests[12] which examined, particularly factors (b) and (c), the results substantiate, in some measure, the short-comings implied under the headings. With an efficient foam-making nozzle at close range to maximise impact, and with a relatively modest impact velocity of 7 m/s, the following major conclusions arose:

1. All heptane fires were extinguished. Many petrol fires were not extinguished. Hence heptane is an unrealistic fuel.
2. Hexane proved more difficult to extinguish than heptane. However hexane-toluene mixtures, though less volatile, proved even more difficult to extinguish than hexane alone. In separate tests 60/40 v/v pentane/toluene appeared to be an acceptable simulant for the commercial petrol used in these tests. (Petrol varies widely in its composition).
3. Some of the foams which had been found to give good results in tests at 4-5 m/s impact velocity, were unable to extinguish petrol fires in the test conditions enumerated above.
4. Foams with a 'protein' base seemed less vulnerable to petrol contamination than those with a 'synthetic' base.

Given that the last conclusion could well be simplistic, the question nevertheless arises as to whether the oleophobic character of some AFFFs can be 'saturated' with consequent scope for high fuel transfers through the foam blanket in severe conditions.

Consequent upon these fire tests we have initiated studies of interfacial relationships with the commercial formulations and laboratory-grade hydrocarbons. These studies are at an early stage but some typical figures are tabulated below:

TABLE 1 Interfacial tensions and spreading coefficients

Foam Agent	Fuel	γ_i mN/m 20°C	50°C	$S_{f/w}$ mN/m 20°C	50°C	$S_{w/f}$ mN/m 20°C	50°C
AFFF I	Heptane	1.8	2.9	- 3.3	0	-0.5	-1.8
	Toluene	2.5	2.1	-14.0	-11.2	8.9	7.2
AFFF II	Heptane	2.8	2.4	- 4.4	- 1.2	-1.2	-2.4
	Toluene	2.6	2.1	-14.7	-11.7	9.0	7.7

In related dynamic studies, foams were applied to heptane and toluene at 6 m/s in conditions indicated by Plates 1, 2 (The heptane used contains nominally 38% n-isomer and has a boiling range of 94-99°C.) The foams used were a fluoroprotein and AFFF II. Results are shown in Table 2.

TABLE 2

Foam	Fuel	Fuel uptake $\left(\frac{\text{vol fuel}}{\text{vol aqueous solution}}\right)$ %
FP	Heptane	14
FP	Toluene	14
AFFF II	Heptane	32
AFFF II	Toluene	27 (see note below)

With fluoroprotein foam, the aqueous and hydrocarbon phases separated well within an hour or two though there was an interfacial scum between the aqueous phase and heptane. There appeared to be some cavitation within the foam ie considerable bubble amalgamation.

By contrast the AFFF showed no cavitation of foam but a substantial degree of emulsification occurred with toluene. Of the replicated samples, one gave substantial separation of fuel phase but the other emulsion was more stable. After 48 hours the latter was centrifuged at 1000 g for 5 minutes. The emulsion still represented 65% of the mixture by volume while (clear) toluene accounted for 29% and a (hazy) aqueous layer for 6%. Since emulsification implies scope for transfer of fuel through aqueous phase, the formation of a stable emulsion is not favourable to fire extinction. No obvious relationship between spreading coefficients and the tendency to emulsify can be discerned. This raises the question of the inclusion of requirements for spreading coefficients in specifications as is currently the case. It is to be hoped that further work will clarify these relationships. It is worth emphasising that even 'small' fire tests are both demanding of resources and pose difficulties of interpretation, with corresponding incentives for establishing alternative means for evaluation of foam agents.

In conclusion, the apparently unsophisticated techniques currently used for formulation, evaluation and application of fire fighting foams arise in part from the interactions covered in the complex and often ill-defined characteristics of fires. Large liquid fuel fires are relatively infrequent yet the consequences may be serious, whether in lives (two large aircraft in collision) or material assets where entire refinery and storage complexes are at risk. In the past, meticulous work on the improvement of protein foams was substantially devalued, almost overnight, by the introduction of fluorosurfactants. At the present time, the industry is pre-occupied with the impending introduction of alcohols etc as octane boosters in motor fuel; water-miscible fuel is foam-destructive. While recent work on foam-fuel interaction may have emphasised the scope for cost-effective improvement, further development along these lines must be seen against the application of limited resources to substantial new developments in the petrochemical industry. However if this paper has stimulated, by omission as well as commission, fresh approaches to fires of liquid fuels, it will lead to greater safety and to conservation of fossil fuels.

References

1. Linacre, E J et al: Ministry of Power Research Reports 171, 179, 182 and 213.
2. Nash, P, Bridge, N W, and Young, R A: Extinction of Fires in Storages of Racked Goods Using High Expansion Foam. Fire Research Note No 857 (1971).
3. Briggs, A A: Protection of Large Flammable Storage Tanks During Repair by Means of Nitrogen-filled High Expansion Foam. J. Hazardous Materials 3 17-21 (1979).

4. French, R J, and Hinkley, P L: The Extinction of Fires in Petrol Storage Tanks by the Base-Injection of Air-Foam. Fire Research Note No 100/1954.
5. Nash, P and Whittle, J: Fighting Fires in Oil Storage Tanks Using Base Injection of Foam. Part I. Fire Technology 14 15-27 (1978).
6. Nicolson, P C and Artman, D D: A Technique for the Evaluation of AFFF Sealing Characteristics. Fire Technology 13 13-20 (1977).
7. Cousin, C S and Briggs, A A: Acceleration of Ignition of Avtur by Unfoamed Aqueous Film Forming Foam Solutions: Preliminary Experiments (In press).
8. For example: UK Patents 1, 070, 289; 1, 258, 299; 1, 270, 661; 1, 270, 838; 1,524, 875; 1, 531, 838; 1, 538, 104; 1, 591, 292.
9. US Military Specification MIL F24385C (1981).
10. Corrie, J G: Measuring the Shear Stress of Fire-Fighting Foams. Fire Research Note No 1055.
11. Briggs, A A: Unpublished work.
12. Briggs, A A and Webb, J S: Testing Concludes Petrol Can Reduce the Effectiveness of Some Foams. Fire 78 45 (Jan 1986).

Acknowledgement

This paper forms part of the work of the Fire Research Station, Building Research Establishment, Department of the Environment. It is contributed by permission of the Director of BRE.

Applications of Surfactants and Polymers in Pesticidal Suspension Concentrates

By Th. F. Tadros
ICI PLC, PLANT PROTECTION DIVISION, JEALOTT'S HILL RESEARCH STATION, BRACKNELL, BERKS. RG12 6EY, UK

In this overview, the application of surfactants and polymers in the preparation of suspension concentrates and maintenance of their physical stability are described. The role of surfactants and polymers in the preparation of suspension concentrates is described by considering their effect on the wetting of the powder by the liquid, breaking of the aggregates and comminution of the resulting particles into smaller units. The role of surfactants and polymers in controlling the colloidal stability is described in terms of the interaction forces between the particles. These interaction forces are classified into electrostatic, van der Waals and steric repulsion. Electrostatic repulsion arises when the particles are dispersed using an ionic surfactant, whereas steric repulsion occurs when using nonionic surfactants and polymers. With polyelectrolytes stability is the result of a combination of electrostatic and steric repulsion. By combining these repulsive forces with the van der Waals attraction, three general energy-distance curves are produced which could be used to describe the state of dispersion. The influence of surfactants and polymers on the rheology of suspensions is described. With sterically stabilised dispersions the viscosity-volume fraction behaviour is similar to that usually observed with hard-sphere dispersions. The same applies with electrostatically stabilised suspensions with particle sizes large relative to the adsorbed layer thickness. However, with flocculated suspensions, or those in which another disperse phase is added for prevention of caking (or claying), the flow behaviour is non-Newtonian and surfactants and polymers affect such flow in various ways. The application of surfactants and polymers for controlling settling of coarse suspensions and prevention of claying is described. The use of polymeric thickeners, which produce non-Newtonian flow and a high residual viscosity are particularly useful in prevention of settling and claying. Finally, the role of surfactants and polymers in prevention of Ostwald ripening (crystal growth) is described by considering the various processes involved.

1. INTRODUCTION

Suspension concentrates, in particular those of the aqueous type, are widely applied in the formulation of many water insoluble pesticides. These systems have attracted considerable attention in recent years in view of the advantages they offer to the user such as ease of handling and application, safety to the operator and environment, and economy. In most cases, these suspension concentrates are prepared by dispersing the powder in an aqueous solution of a dispersing agent, usually a surface active agent or polymer followed by a wet milling process resulting in a size distribution in the range 0.1 - 5 m. As we will see later, the dispersing agent aids the wetting of the powder, enables the breaking of aggregates and agglomerates and subsequent subdivision of the large particles into smaller units. In addition, the agent which becomes adsorbed on the particle surface should ensure colloidal stability of the dispersion (see below).

In this overview, which is by no means exhaustive, some applications of surfactants and polymers in the preparation of pesticidal suspension concentrates and the maintainance of their physical stability will be described. Three main processes need to be controlled in a suspension(1). The first and most important is the control of particle aggregation which determines the final properties of the suspension such as its flow characteristics, its behaviour on application, and its state on standing. Particle aggregation is determined by the interaction forces that exist between them. These forces may be classified into electrostatic, van der Waals and steric forces and the different types of interaction forces will be described in the next section with particular emphasis on the role of surfactants and polymers. The effect of surfactants and polymers on the flow behaviour (rheology) of a suspension will also be discussed. The second process that needs to be controlled is that of "caking" or "claying" which occurs with coarse suspensions that are colloidally stable. Application of surfactants and polymers in controlling the settling of suspensions will be briefly described. Finally with suspensions with a wide distribution of particle sizes, a process in which there is a shift in distribution towards larger sizes (as a result of the preferential dissolution of smaller particles and subsequent growth of the larger ones), i.e. the so-called Ostwald ripening process, needs to be controlled. The role of surfactants and polymers in such a process will also be discussed in subsequent sections.

2. APPLICATION OF SURFACTANTS AND POLYMERS IN PREPARATION OF SUSPENSION CONCENTRATES

As mentioned in the Introduction, pesticidal suspension concentrates are prepared by dispersing the powder into an aqueous solution of a dispersing agent followed by a wet milling process. The role of surfactants and polymers in the dispersion process can be seen from a consideration of the stages involved (2). Three steps have been considered by Parfitt (3,4): wetting of the solid powder by the liquid; breaking of the aggregates; and comminution of the resulting particles into smaller units. It is obvious that the particles formed should be maintained in the dispersed state during the above mentioned processes.

Generally, the particles in a dry powder either form aggregates, in which the particles are joined together at their crystal surfaces, or agglomerates, in which the particles touch only at edges or corners, forming looser, more open structures. For complete wetting of the powder, it is essential to replace all the air (which may be entrapped between the particles, present in pores, and/or adsorbed) and other contaminants by the liquid. In other words, the wetting of the powder by a liquid is the process in which the solid-air interface is replaced by a solid-liquid interface. It is essential to wet the external and internal surfaces and such wetting is described in terms of the equilibrium contact angle of the liquid on the solid surface and the appropriate surface tensions. Three types of wetting processes may be distinguished: adhesional wetting, W_a, when the surface of the solid is brought into contact with the liquid, immersional wetting, W_i, when the solid particles are immersed in the liquid and spreading wetting, W_s, when the liquid spreads over the surface of the partially immersed particles. Assuming the particles consist of cubes with unit length, the work of dispersion W_d (which is the sum of W_a, W_i and W_s) is given by the equation,

$$W_d = -6 \gamma_{LV} \cos\theta \qquad (1)$$

where γ_{LV} is the surface tension of the liquid (liquid/vapour interface) and θ is the contact angle of the liquid on the surface of the solid.

Thus, the wetting of a solid by a liquid depends on two measurable parameters, γ_{LV} and θ. In practice, since surface active agents are added in sufficient quantities, is often very close to zero and spontaneous dispersion of a powder in a surfactant solution is the rule rather than the exception. The addition of the surface active agent also causes a reduction in γ_{LV} and if adsorbed on the surface of the solid also reduces γ_{SL}. Both effects result in enhanced wetting. Morever, wetting of the internal surface, i.e. penetration of the liquid into channels and inside the agglomerates can also be brought about by the surfactant. This can be understood from the capillary phenomenon in which the pressure, P, required to force a liquid through a capillary is given by the equation,

$$P = (-2 \gamma_{LV} \cos\theta)/r = -2(\gamma_{SV} - \gamma_{SL})/r \qquad (2)$$

where r is the capillary radius. Penetration is spontaneous as long as $\theta < 90^{\circ}$ and equation (2) shows that for spontaneous penetration γ_{SL} must be made as small as possible, which is achieved by surfactant adsorption.

Another role played by the surfactant is in breaking aggregates and agglomerates, which is enhanced by penetration of the liquid into the channels (as a result of the wetting phenomena described above) as this provides a pressure to bring about the disintegration of the agglomerates. This is aided by the use of high speed stirrers. Finally the surfactants may also play a role in the comminution process which involves rupturing of chemical bonds to create new surfaces. It has been argued that reduction of surface energy may facilitate comminution by retarding the rejoining of the ruptured surface and by preventing particle agglomeration. The role of surface active agents in the comminution process has been considered by Rehbinder and coworkers (5, 6) who claimed that adsorption of surface active agents at structure defects in the surface (which results in a reduction of the interfacial energy at the boundary between solid and medium) should aid the process of deformation or destruction of that surface.

3. ROLE OF SURFACTANTS AND POLYMERS IN CONTROLLING COLLOIDAL STABILITY

As mentioned in the Introduction, control of particle aggregation is determined by the interaction forces that exist between them, namely electrostatic, van der Waals and steric forces. Below, a brief description of these three main interaction forces and the role of surfactants and polymers will be given. This is followed by a discussion of the total interaction and description of the state of the suspension.

3.1 Electrostatic Interactions: These result when using ionic surfactants of the anionic, cationic and zwitterionic type or polyelectrolytes. As a result of their adsorption on the particle surface, an electrical double layer develops at the solid/solution interface. The double layer consists of the surface charge resulting from the adsorbed surfactant ion or polyelectrolyte (eg. SO_4^- charges when using sodium dodecyl sulphate, SO_3^- charges when using lignosulphonates, or N^+ charges when using tetra-alkyl ammonium salts) and a compensating diffuse layer of ions that extends into bulk solution to an extent depending on the total electrolyte concentration.

When two particles, each carrying a diffuse layer of counterions, approach each other to a distance that is comparable to twice the double layer thickness, repulsion results from overlap of the diffuse layers. For spherical particles, the repulsive energy, G_E, is given by the approximate Derjaguin expression (7) when the range of interaction is small compared to particle radius,

$$G_E = 2 \pi a \varepsilon_r \varepsilon_o \psi_o^2 \ln [1 + \exp(-\kappa h)] \qquad (3)$$

where ε_r is the dielectric constant of the medium, ε_o is the permittivity of free space, ψ_o is the surface potential (which is approximately equal to the zeta potential when there is no specific adsorption of counter or co-ions), h is the surface-surface separation and κ is the reciprocal of the double layer "thickness", that is given by the following expression for 1:1 electrolytes,

$$\kappa = \left(\frac{2 z^2 e^2 C}{\varepsilon_r \varepsilon_o kT} \right)^{1/2} \qquad (4)$$

where z is the valency of the counterions, e the electronic charge, C the electrolyte concentration, k the Boltzmann constant and T the absolute temperature.

Equation (3) is only valid for low ψ_o value ($z \psi_o < 25$ mV) when $\kappa a \gg 1$ and $\kappa h > 1$. G_E depends on ψ_o (which depends on the number of surfactant ions adsorbed per unit area), particle radius a and κ (determined by electrolyte concentration). The higher ψ_o the higher the repulsion and the higher κ, the smaller G_E. This shows that for increasing repulsion between particles, a high adsorption density of surfactant ions and a low electrolyte concentration are required.

3.2 <u>van der Waals Interaction</u>: This arises from the dispersion (London) force, which is the result of charge fluctuations within an atom or a molecule associated with the motion of its electrons.. This attraction energy between single atoms or molecules is of short range, being proportional to the sixth power of the distance of separation between them. However, the total attraction energy between agglomerates of atoms or molecules (as in the case of suspension particles) is of a much longer range (being approximately equal to the sum of all interaction energies between atoms or molecules within one particle with that of all other atoms or molecules in the second particle) which for two spherical particles is given by the approximate Hamaker equation (8),

$$G_A = -\frac{A\ a}{12\ h} \qquad (5)$$

where A_{12} is the effective Hamaker constant for particles 1, with Hamaker constant A_{11}, immersed in a medium 2, with Hamaker constant A_{22},

$$A_{22} = (A_{11}^{1/2} - A_{22}^{1/2})^2 \qquad (6)$$

The Hamaker constant of any material A_{ii} is related to the London dispersion constant β_{ii} and the number density of atoms or molecules, q_i by the equation,

$$A_{ii} = \pi^2\ q_i^2\ \beta_{ii} \qquad (7)$$

β_{ii} is related to the polarizability of the atoms or molecules concerned.

In the presence of an adsorbed surfactant or polymer layer, the net Hamaker constant is related to that of the particle A_{11}, the medium A_{22} and that of the adsorbed layer A_{33} (9),

$$G_A = -\frac{1}{12}\left[(A_{22}^{1/2} - A_{33}^{1/2})^2\ \frac{a+\delta}{a} + (A_{33}^{1/2} - A_{11}^{1/2})^2\ \frac{a}{(\Delta + 2\delta)} + \frac{4\ a\ (A_{22}^{1/2} - A_{33}^{1/2})\ (A_{33}^{1/2} - A_{11}^{1/2})\ (a+\delta)}{(\Delta+\delta)\ (2\ a+\delta)}\right] \qquad (8)$$

where δ is the thickness of adsorbed surfactant or polymer layer and Δ is the surface-surface separation of the layers. It is clear from equation (8) that the effect of surfactant or polymer layer depends on the relative magnitudes of A_{11}, A_{22} and A_{33} (10). In most cases $A_{11} > A_{33} > A_{22}$ and hence the presence of a surfactant or polymer layer leads to a reduction in G_A at any given distance of separation between the particle surfaces h.

3.3 Steric Interactions: These result from interaction between adsorbed surfactant, polymer or polyelectrolyte layers (11). This is, for example, the case when using non-ionic surfactants of the ethoxylate type, i.e. $R-(CH_2CH_2O)_n$ OH containing several ethylene oxide units or macromolecular surfactants of the A-B or A-B-A block type or graft type B A_n, e.g. Pluronics,

$$HO - (CH_2CH_2O)_x - (CH_2\underset{\substack{|\\ CH_3}}{CH} - O)_y - (CH_2 - CH_2O)_x - OH$$

When such molecules are used for the preparation of pesticidal suspension concentrates, the hydrophobic moieties (R - or polypropylene oxide in the above case) become attached to the hydrophobic particle surface, leaving the hydrophilic moieties of poly(ethylene oxide) dangling in solution. When two particles with such adsorbed layers approach each other to a separation distance whereby the layers interfere, i.e. when $h < 2\delta$, steric interactions arise from two main effects. The first effect, usually referred to as the mixing effect, results from a build-up of segment concentration in the interaction zone (overlap region). In this region, the chemical potential of the polymer chains is greater than that in the rest of the adsorbed layer. This gives a free energy of interaction term, G_M,

$$G_E = \frac{4\pi\phi_2^2}{3v_1}\left(\frac{1}{2} - \chi\right)\left(\delta - \frac{h}{2}\right)^2\left(3a + 2\delta + \frac{h}{2}\right) \qquad (9)$$

Equation (9) was derived by Fischer (12) assuming a uniform segment density in the adsorbed layer (clearly an invalid assumption with polymer molecules). In equation (9), ϕ_2 is the volume fraction of adsorbed surfactant or polymer, V is the molar volume of the solvent (water in the case of aqueous suspensions) and χ the chain-solvent interaction parameter (Flory-Huggins interaction (13)). From equation (9) G_M is positive when $\chi < 0.5$, i.e. the surfactant or polymer chains are in good solvent conditions. This is the case with the above examples, where the stabilising moiety consists of poly (ethylene oxides) at ordinary temperatures. These chains are soluble in water at ordinary temperatures as a result of hydrogen bonding between the ether oxygen of ethylene oxide and water. However, if such a hydrogen bonding mechanism is broken down, as for example, is the case in the presence of high electrolyte concentration and/or increase of temperature, the chains lose their solvation shell and chain-chain interaction becomes favourable. In this case $\chi > 0.5$ and G_M changes sign, i.e. it becomes negative so that the mixing term becomes attractive. Clearly, there is one well defined point at which $\chi = 0.5$ and $G_M = 0$. This is called the θ- point of the chain, which can be defined as the critical electrolyte flocculation concentration (CFC) at constant temperature or critical flocculation temperature (CFT) at constant electrolyte concentration.

The second effect in steric interaction results from compression of chains as a result of closer approach. This is usually the case when the distance of separation between the particle surfaces is smaller than one adsorbed layer thickness. Under these conditions, the chains lose configurational entropy (14) since the volume available for the chain becomes restricted. The effect is referred to as the volume restriction G_{VR} or elastic, G_{el}, effect. For simple surfactant molecules at low coverage, a simple expression for G_{VR} has been derived by Mackor and van der Waals (15),

$$G_{VR} = N_S \ kT \ \theta_\infty [1 - (h/\delta)] \tag{10}$$

where N_S is the number of adsorbed chains per unit area and θ_∞ is the fraction of surface covered by an adsorbed layer with surfaces at infinite separation. Clearly, more elaborate expressions are required for G_M and G_{VR}, when dealing with high surface coverages and especially with macromolecules, where the adsorbed layer is not uniform and a large number of segments are present in the chain (16). It should be mentioned that, unlike G_M, G_{VR} is always positive.

The steric interaction, G_S, is given by the sum of G_M and G_{VR}, i.e.,

$$G_S = G_M + G_{VR} \tag{11}$$

G_M operates as soon as $h < 2\delta$, whereas G_{VR} operates at further overlap and becomes significant when $h < \delta$.

3.4 Total Interaction and Description of the State of the Suspension Concentrate: The total interaction energy of particles containing adsorbed surfactant, polymer or polyelectrolyte chains is given by the general expression

$$G_T = G_E + G_S + G_A \tag{12}$$

Three general cases may be considered, the first where there is little contribution from G_S, i.e. the case when using ionic surfactants. In this situation, G_T is given by the sum of G_E and G_A in the well known theory of Derjaguin, Landau, Verwey and Overbeek (17) (DLVO theory). The form of energy-distance curve is illustrated in Fig.1a. The energy-distance curves is characterised by three main parameters: G_{max} $(G_{min})_{primary}$ and $(G_{min})_{sec}$. the magnitudes of which are determined by the surface potential ψ_o (or the

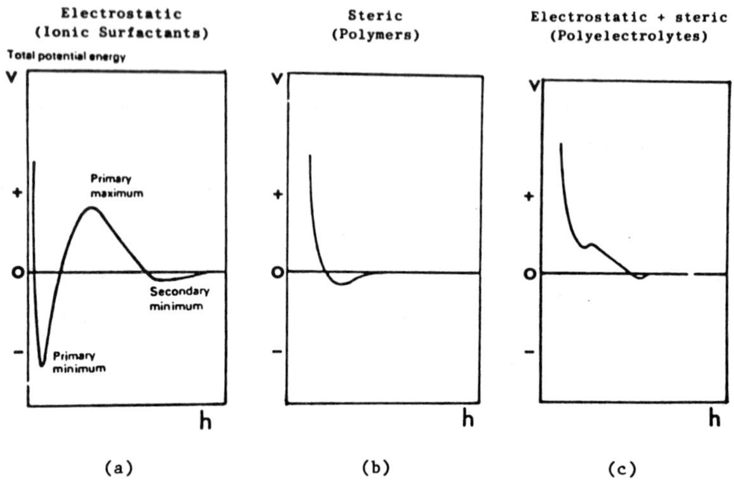

Fig. 1 - (a) Total energy-distance curve according to the DLVO theory; (b) Total energy-distance curve for sterically stabilised systems (non-ionic surfactants or polymers); (c) Total energy-distance curve for systems stabilised by polyelectrolytes.

measured ζ- potential), electrolyte concentration C, particle size a and Hamaker constant A . Depending on the relative magnitudes of these parameters, various states of the suspension are reached. If G_{max} is high ($\gg kT$) i.e. when ψ_o or ζ is high and C is low, a deflocculated suspension is produced. As we will see later, with such deflocculated suspensions, dilatant clays are produced as a result of settling of the coarse suspension particles to the bottom of the container. When G_{max} is small (<5 - 10 kT, or absent), the suspension becomes coagulated i.e. it is in the primary minimum. On the other hand, when G_{max} is reasonably large ($>$ 10 kT) and $(G_{min})_{sec}$ is fairly deep, weak flocculation occurs and the particles are held by a relatively weak attraction force at distances around the location of $(G_{min})_{sec}$. This is usually the case with coarse suspensions (particularly when the particles are anisometric) when the stability is reduced by the addition of electrolyte levels that are lower than required for coagulation (18). Under these conditions the suspension may form a three-dimensional gel network structure, which can be fluidised by shaking. This state may be used for prevention of caking (18).

The second case to consider is where the repulsive forces are dominated by the steric interactions, i.e. $G_S >> G_E$. In this case the total interaction is given by the sum of G_S and G_A (19). The form of the energy-distance curve is illustrated in Fig. 1b. As is clear, only one minimum G_{min}, is present whose depth depends on the adsorbed layer thickness δ, the particle radius a and the Hamaker constant A. For systems with thick adsorbed layers, as for example the case with nonionic polymers, G_{min} is shallow (of the order of one kT unit or less) and in this case the suspension approaches thermodynamic stability at low particle number concentrations. However, even with such low G_{min} values, some weak flocculation occurs above a critical volume fraction ϕ. This difference in behaviour between dilute and concentrated suspensions can well be understood from the balance between energy and entropy of flocculation. The total free energy of flocculation is given by the expression

$$\Delta G_{floc} = \Delta G_h - T\Delta S^{conf} \qquad (13)$$

ΔG_h is the term associated with interaction which is related to G_{min}, whereas ΔS^{conf} is the loss in configurational entropy of the particles on flocculation. With dilute suspensions, ΔS^{conf} is relatively large and negative (since the particles lose configurational entropy on flocculation) and when ΔG_h is relatively small, the $-T\Delta S^{conf}$ term, which is now positive, exceeds the small negative value of ΔG_h so that ΔG_{floc} is positive and the process is unfavourable. This is the origin of thermodynamic stability with such dilute systems. However, with increase in particle number concentration, the net value of $T\Delta S^{conf}$ decreases since the configurational entropy of the particles before flocculation is reduced with increase in particle number concentration. At a critical ϕ value, the net positive $T\Delta S^{conf}$ term balances the ΔG_h terms and ΔG_{floc} becomes zero. Thus a critical ϕ value may be defined at which flocculation begins. Above ϕ_{crit}, weak flocculation increases in magnitude with increase in ϕ .

It is clear from the above discussion that ϕ_{crit} depends on the magnitude of ΔG_h. The larger ΔG_h, i.e. the deeper the minimum, the smaller ϕ_{crit} at which flocculation begins. Apart from this mechanism of weak flocculation, instability in sterically stabilised suspension may be induced by reduction in solvency of the medium to the chains (see above). This is schematically illustrated in Fig. 2, which shows the effect of changing the χ parameter from a value < 0.5 to a value > 0.5. When $\chi > 0.5$, G_M becomes negative and a deep minimum is produced in the G - h curve. This condition results in catastrophic flocculation of the suspension (usually referred to as incipient flocculation). As mentioned above, changes in the χ-parameter with ethoxylated surfactants or polymers are produced by temperature changes and/or addition of a non-solvent for the chain, eg. high concentrations of electrolyte.

The third case to consider is where both G_E and G_S are significant such that the total energy of interaction is given by the sum of G_E, G_S and G_A. This is particularly the case with polyelectrolytes and the form of the energy-distance curve is illustrated in Fig. 1c. Examples of polyelectrolyte type systems in pesticidal suspensions concentrates are best illustrated by the naphthalene-formaldehyde sulphonated condensates and lignosulphonates. With naphthalene-formaldehyde sulphonated condensates, oligimers with 2-10 units are produced. Studies by Garvey and Tadros (19) have shown that steric interactions are significant particularly with the high molecular weight fractions. Since these chains are strongly charged, double layer interaction is also significant. The other most commonly used polyelectrolyte system for the stabilisation of pesticidal suspension concentrates is that of lignosulphonates. Studies using both practical (20) and model suspensions of polystyrene latex (21) have clearly indicated that both electrostatic and steric forces are operative in such systems.

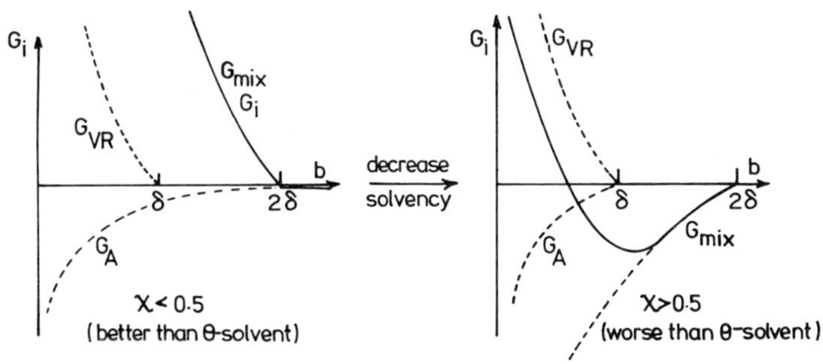

Fig. 2 Influence of change of solvency on the energy-distance curve.

4. INFLUENCE OF SURFACTANTS AND POLYMERS ON THE RHEOLOGY OF SUSPENSIONS

Suspension concentrates when colloidally stable may show Newtonian behaviour up to high volume fractions. This is particularly the case with sterically stabilised suspensions in which the adsorbed layer thickness is relatively small compared with the particle radius. Under these condtions, the suspension will behave like a hard-sphere system with very small interaction until the layers begin to interfere. This occurs at values approaching close-packing (~ 0.6 for random packing and ~ 0.7 for hexagonal packing). As an illustration, the viscosity-volume fraction curve for a pesticidal suspension concentrate (ethirimol) is shown in Fig. 3. This curve shows a rapid increase in η above $\phi = 0.6$. In this case it may be possible to fit the data using the hard-sphere model suggested by Dougherty and Krieger (22)

$$\eta_r = [1 - (\phi / \phi_p)]^{-[\eta]\,\phi_p} \qquad (14)$$

where $[\eta]$ is the intrinsic viscosity which is theoretically equal to 2.5 and ϕ_p is the so-called packing fraction. Results for the calculation of η_r using equation (14) with $[\eta] = 2.5$ and $\phi_p = 0.7$ are shown in Fig. 3. It can be seen that the calculated η_r values are lower than the experimental values, indicating that the suspension does not behave as a perfect hard-sphere system and/or due to the fact that at high ϕ values, a residual attraction is operative and this needs to be taken into account. It should also be mentioned that the above a critical ϕ value the suspension shows viscoelastic behaviour (with the elastic component being larger than the viscous component) due to the interaction of the adsorbed layers (23).

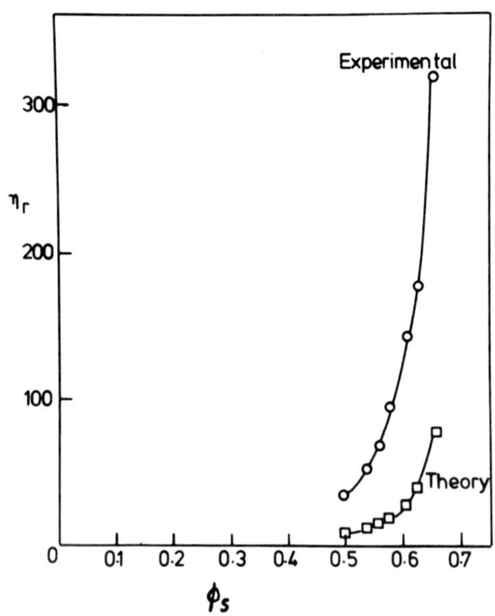

Fig. 3 Viscosity - volume fraction curve for a sterically stabilised pesticidal suspension.

Electrostatically stabilised suspensions also show Newtonian behaviour to relatively high ϕ values. This is again due to the fact that the double layer thickness is relatively small compared to particle radius particularly when the electrolyte level is not too small. Results of η_r versus ϕ for polystyrene latex suspensions stabilised by lignosulphonates showed that a steep rise at ϕ values greater than ~ 0.6. However, with suspensions consisting of small particle size (submicron) and extended double layers, i.e. at low electrolyte concentrations, viscoelasticity is exhibited at relatively low ϕ values as a result of the interaction of the double layers (24).

Flocculated suspension concentrates, or those in which another disperse phase such as sodium montmorillonite or oxide is added to build up a three-dimensional structure for prevention of claying or caking, show non-Newtonian behaviour. The most common flow behaviour for such systems is pseudoplastic and this is illustrated in Fig. 4 which shows the shear stress (σ) shear rate ($\dot{\gamma}$) relationship. The flow curve is characterised by three main parameters: (a) $\dot{\gamma}_{crit}$, the shear rate above which the flow curve becomes linear. $\dot{\gamma}_{crit}$ is the value above which viscous stresses rupture all flocs

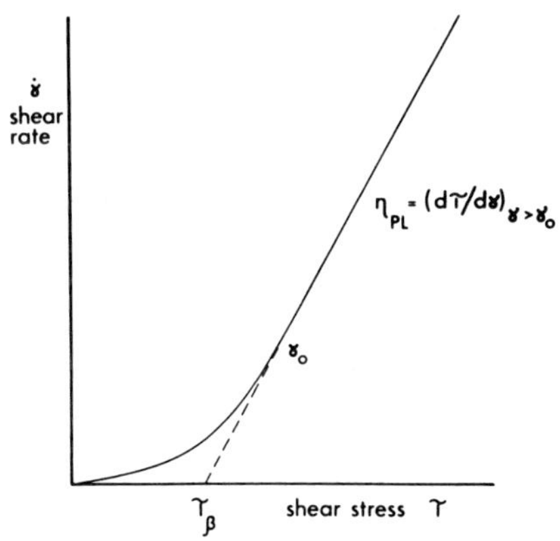

Fig. 4 Flow curve of a flocculated or "structured" suspension.

formed; (b) η_{pl} (= $(d\sigma/d\dot{\gamma})_{\dot{\gamma}>}$ $\dot{\gamma}_{crit}$ i.e. the gradient of the linear shear stress-shear rate curve. The pseudoplastic viscosity arises from purely hydrodynamic effects; (c) σ_{β} the value of the stress obtained by extrapolation of the linear portion of the flow curve to $\dot{\gamma}$ = 0. The residual stress arises from the residual effect of the interaction potential.

Surfactants and polymers both ionic and nonionic influence the rheology of flocculated or structured suspension concentrates in various ways. For example, the surfactant micelles will contribute to the rheology of the suspensions, particularly when such units are large and anisometric. This is particularly the case at high surfactant concentrations (which are sometimes used for enhancement of biological efficacy) whereby rod-like micelles and liquid crystalline phases are formed. Polymers can affect the rheology by contributing to the net-interaction e.g. polymer bridging. Moreover, high molecular weight polymers are often added to control the settling and prevent formation of dilatant sediments (see below). The level added depends on the molecular weight of the polymer and in all cases the macromolecules should form a viscoelastic network resulting from overlap of the polymer coils. This is reflected in the net rheology of the suspension which will also be viscoelastic with sufficient elasticity to prevent settling of the coarse suspension particles (see below).

5. APPLICATION OF SURFACTANTS AND POLYMERS FOR CONTROLLING SETTLING AND PREVENTION OF CLAYING

As mentioned in the Introduction, settling of coarse suspensions which are colloidally stable leads to the formation of dilatant sediments. The repulsive forces necessary to ensure colloidal stability enable the particles to move past each other in the sediment with the resulting of formation of a very compact layer. As a consequence of the dense packing and small spaces between the particles such "clays" are difficult to redisperse. It is possible, at least in principle, to prevent the formation of compact layers by controlling the interparticle interaction. For example, with suspensions containing ionic surfactants, compact sediments may be prevented by secondary minimum flocculation. With systems containing nonionic or macromolecular surfactants, compact sediments may be prevented by controlling the adsorbed layer thickness. As shown in Fig. 1b, for a given system, G_{min} depends on the magnitude of δ. Thus, by reducing δ, G_{min} becomes deeper reaching values that are sufficient to induce weak flocculation and prevent formation of dilatant sediments.

Although the above ideas are sound in principle, their practical application is not easy since a number of parameters need to be controlled simultaneously. For this reason, several other more practical methods are used for prevention of claying. For example, high concentrations of surfactants may be applied. As mentioned above, the concentration must be sufficiently high to form rod-shaped micelles which produce viscoelastic behaviour. Alternatively, a high molecular weight polymer may be added at much lower concentrations. Examples of such polymers (sometimes referred to as thickeners) are hydroxyethyl cellulose, polysaccharides (such as Xanthan gum) and polyacrylamides. These high molecular weight materials produce viscoelasticity in the continuous medium and give high viscosities at low shear rates. As an illustration, the results obtained by Buscall et al (25) using ethyl hydroxyethyl cellulose are worth mentioning. The authors measured the viscosity of this polymer at various concentrations (0.22 - 1.48% w/v) at various shear stresses reaching very low values. The results are shown in Fig. 5 which clearly demonstrates the shear thinning behaviour of these polymer solutions. Below a critical shear stress the viscous behaviour is Newtonian and the critical shear stress is in the region of 0.2 Nm^{-2}. From the curves shown in Fig. 5, the limiting viscosity at low shear stress, i.e. in the

Newtonian region can be obtained. This limiting value of the viscosity at vanishingly small shear stres (or shear rate) is termed the residual viscosity.

The settling rate of a 5% coarse (1.55 μm radius) latex was measured in various ethyl hydroxyethyl cellulose solutions covering the above range. A plot of u/a^2 (where u is the sedimentation rate versus $\eta_{(o)}$ is shown in Fig. 6. A linear curve with a slope equal to -1 is obtained, indicating the correlation between settling rate and the zero shear viscosity. Thus, with

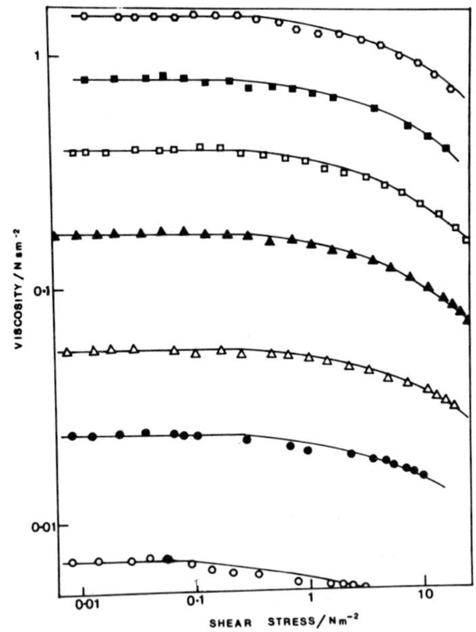

Fig. 5 Viscosity against shear stress for various concentrations (% w/v) of ethyl hydroxy eythyl cellulose in water: 0, 0.22; ●, 0.43; △ 0.65; ▲ 0.86; □ 1.08; ■ 1.3; ○ 1.48.

the present model system of coarse polystyrene latex suspensions, good long term stability with respect to settling would occur with a zero shear viscosity of 10 Pas. However, much higher values are required with many practical suspensions, when the density of the particles is much higher than that of the medium. In the present model suspension of polystyrene latex, the density of the particles is 1.05, whereas with many pesticidal suspension concentrates densities of 2 or even higher are encountered. Moreover, with many such practical suspensions, a polymer thickener may not be sufficient to prevent claying, since the concentration of the polymer cannot be increased to high levels, as this results in deterioration of spontaneity of dispersion on dilution. In many cases a mixture of a finely divided solid such as sodium

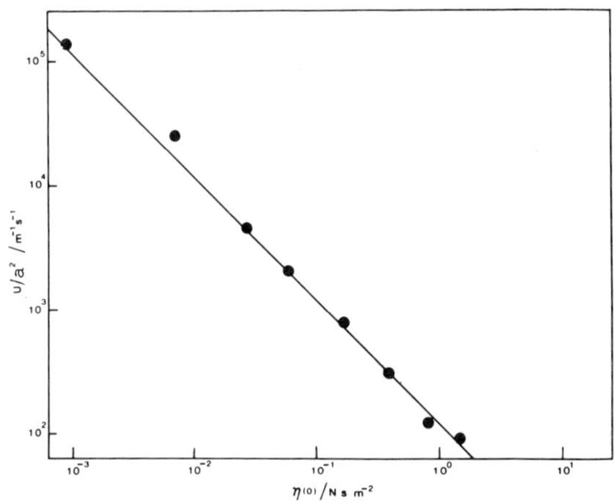

Fig. 6 u/a^2 versus $\eta_{(o)}$ for a latex suspension (ϕ = 0.05) in aqueous solutions of ethyl hydroxyethyl cellulose.

montmorillonite or silica with a polymer such as hydroxyethyl cellulose or Xanthan gum is used. In such cases, the mechanism of prevention of claying is probably due to the presence of a viscoelastic gel network in the continuous phase. Under these conditions, the correlation between settling and claying and the rheological characteristics is not straightforward and various parameters such as yield value, zero shear viscosity, complex, storage and loss moduli need to be determined for full evaluation of the rheology. Such rheological measurements and the interpretation of their results are beyond the scope of the present overview.

6. ROLE OF SURFACTANTS AND POLYMERS IN OSTWALD RIPENING

Ostwald ripening or crystal growth is the result of the difference in solubility between small and large particles (26)

$$\frac{RT}{M} \ln \frac{S_2}{S_1} = \frac{2\gamma_s}{\rho} \left(\frac{1}{r_1} - \frac{1}{r_2}\right) \qquad (15)$$

where S_1 and S_2 are the solubilities of crystals of radii r_1 and r_2 respectively, γ_s is the specific surface energy, ρ is the density and M is the molecular weight of the soluble molecules, R is the gas constant and T the absolute temperature. Thus, with time the smaller crystals dissolve preferentially and the solute molecules so created become deposited on the large crystals. The growth of crystals may create undesirable changes in the suspension concentrate. As a result of increase of particle size, the settling of particles may be accelerated, leading to caking and cementing of particles in a sediment. Moreover, increase in particle size may lead to a reduction in biological efficacy.

Surfactants and polymers play a major role in the above process of Ostwald ripening, which can be understood if one considers the main processes involved. These consist of two main steps (27, 28) namely the diffusion of solute to the surface of the growing crystal which is followed by incorporation of molecules in the structure of the crystal lattice. A surfactant or polymer may affect the rate of crystal growth by modifying one or both of these steps. In the first place, the presence of surfactant or polymer influences the diffusion process in several ways by affecting the transport away from the boundary layer (29), although its addition is unlikely to change the rate of dissolution proper (passage from the solid to the dissolved state in the immediately adjacent layer). The dissolution rate can thus be greatly accelerated (30). On the other hand, since surfactant and some polymers form micelles that are capable of solubilising the solute, the diffusion coefficient D of the solute in a micellar solution will be reduced. However, as a result of solubilisation the concentration gradient $\partial C/\partial x$ will be greatly increased in direct proportion to solubilisation (31). Since the flux, J, of the solute in mol $cm^{-2}s^{-1}$ is governed by the product of D and $\partial C/\partial x$, i.e.

$$J = - D \, (\partial C/ \partial x) \qquad (16)$$

the net effect depends on the relative effect on D and $\partial C/\partial x$. Since D is inversely proportional to the radius of the diffusion particle, it is only reduced by about one order of magnitude (corresponding to a micelle having a volume 1000 times larger than that of the solubilizate). On the other hand $\partial C/\partial x$ can be increased by several orders of magnitude as a result of solubilisation. Thus, the net effect will be an increase in the flux, when micelles are present. Solubilisation by micelles may, under certain conditions, enhance transport by diffusion, thus enhancing crystal growth (32).

Surfactants and polymers can also influence the second step, namely surface deposition and incorporation in the crystal lattice (33). Adsorption of a surfactant or a polymer on the surface of a crystal can drastically change the specific surface energy and make it, effectively, inaccessible to the solute. Moreover, if the surfactant or polymer is preferentially adsorbed at one or more faces of the crystal (for example, by electrostatic attraction between, say, a highly negative face of the crystal and a cationic surfactant), surface nucleation is no longer possible at this particular face (5). Growth will then take place at the remaining faces which are either bare or incompletely covered with the surfactant or polymer. This results in a change in crystal habit.

From the above discussion, it can be seen that surfactants or polymers, if properly chosen, may be used for crystal growth inhibition in suspension concentrates (32) and control of habit formation. However, it should be mentioned that crystal growth inhibition is rather specific and it is not possible to draw general rules.

REFERENCES

1. Tadros, Th.F., Adv. Colloid and Interface Science 12, 141 (1980).
2. Patton, P.C., "Paint Flow and Pigment Dispersion", Interscience, New York (1964).
3. Parfitt, G.D., J. Oil Chem. Assoc. 50, 822 (1967).
4. Parfitt, G.D., "Dispersion of Powders in Liquids", Applied Science Publishers 3rd Ed. (1981).
5. Rehbinder, P.A. and Likhtman, V.J., Proc. 2nd Int. Conference Surface Activity, Butterworth (London) No.3 (1975).
6. Rehbinder, P.A., Colloid J. USSR 20, 493 (1958).
7. Derjaguin, B.V., Kolloid Z. 69, 155 (1939); Acta Physica Chemica USSR 10, 313 (1939).
8. Hamaker, H.C., Physica (Utrecht) 4, 1058 (1937).
9. Vold, M.J., J. Colloid Sci. 16, 1 (1961).
10. Vincent, B,, J. Colloid Interface Sci. 42, 270 (1973).
11. Tadros, Th.F. in "Effect of Polymers on Dispersion Properties", Academic Press, London (1982).
12. Fischer, E.W., Kolloid Z. 160, 120 (1958).
13. Flory, P.J., "Principles of Polymer Chemistry" Cornell Univ. Press (1953).
14. Mackor, E.L., J. Colloid. Sci. 6, 492 (1951).

15. Mackor, E.L. and van der Waals, J.H., J. Colloid Sci. 7, 535 (1952).

16. Napper, D.H., "Polymeric Stabilization of Colloidal Dispersions", Academic Press, London (1983).

17. Verwey, G.J.W. and Overbeek, J.Th.G., "Theory of Stability of Lyophobic Colloids" Elsevier Publ. Co., Amsterdam (1948).

18. Tadros, Th. F., Colloids and Surfaces, in press (1986).

19. Garvey, M.J. and Tadros, Th.F., Proc. VI International Congress Surface Activity II (2), 715 (1973).

20. Tadros, Th.F., Advances in Pesticide Sci., Pergammon Press, Part 3, 820 (1979).

21. Tadros, Th.F., Colloid and Polymer Sci., 258, 439 (1980); Heath, D and Tadros, Th.F., Colloid and Polymer Sci., 261, 49 (1983).

22. Krieger, I.M., Adv. Colloid Interface Sci. 3, 111 (1972).

23. Tadros, Th.F. and Zsednai, A., to be published.

24. Buscall, R, Goodwin, J.W., Hawkins, M.W. and Ottewill, R.H., J. Chem. Soc. Faraday I, 78, 2873, 2889 (1982).

25. Buscall R, Goodwin, J.W., Ottewill, R.H. and Tadros, Th.F., J. Colloid Interface Sci. 85, 78 (1982).

26. Ostwald, W, J. Phys. Chem. 34, 493 (1900).

27. Tape, G and Kahlweit, M., J. Phys. Chem. 30, 90 (1961).

28. Nancollas, G. and Purdie, N., Quart. Rev. 18, 1 (1964).

29. Merrie, R.C. and McBain, J.W., J. Phys. Chem. 46, 10 (1966).

30. Bates, T.R., Gibaldi, M. and Konig, J.L., Nature 210, 1321 (1966).

31. Mysel, K.J., Adv. Chem. Ser. (Amer. Chem. Soc.) 86, 24 (1969).

32. Tadros, Th.F., in "Particle Growth in Suspensions", Ed. Smith, A.L. Academic Press, London (1977).

33. Moillet, J.L., Collie, B. and Black, W., "Surface Activity", E. and F.N. Span, London (1961), Chapter 5.

Surfactants in Enhanced Oil Recovery

By T. F. Ling, H. K. Lee and D. O. Shah*
CENTRE OF SURFACE SCIENCE AND ENGINEERING, UNIVERSITY OF FLORIDA, 425 CHEMICAL ENGINEERING BUILDING, GAINESVILLE, FLORIDA 32611, USA

ABSTRACT

Since last several decades various methods have been suggested and tested for the improvement of enhanced oil recovery. The basic objectives of these methods are how to improve sweep efficiency and displacement efficiency or both simultaneously. This paper will review some fundamental aspects of the surfactant-polymer flooding and foam flooding processes, which involve the application of surfactants.

For surfactant-polymer flooding, the major parameters are:

1. Ultralow interfacial tension
2. Interfacial viscosity of oil/brine interface
3. Optimal salinity
4. Surfactant-polymer incompatibility
5. Coalescence and the propagation of oil ganglia
6. Surface charge density at oil/brine and rock/fluid interfaces
7. Wettability at rock/fluid interface

For foam flooding, interfacial properties such as surface tension, foaminess, foam quality, surface viscosity, bubble size distribution etc. of foams have been discussed. These are further correlated with fluid displacement in porous media. The observations of microscopic foam behavior in porous micromodel are also reported.

Some novel concepts for surfactant-polymer flooding and foam flooding processes have been proposed including the use of nonionic surfactants to form the alcohol-free microemulsions and the injection of foam for the mobility control for heavy oil recovery using surfactant-polymer flooding.

INTRODUCTION

When a new oil reservoir is discovered, the reservoir usually possesses the natural pressure to move the oil to the production wells by expansion of volatile components. This phase is called the primary oil recovery. When the natural pressure decreases, it is necessary to increase the pressure by injection of water. The injection of water, so called the secondary oil recovery or water flooding, is stopped when the production cost becomes significant compared to the value of the produced oil. The total oil recovery by the primary and secondary oil recovery is usually less than 40% of the original oil in place. Therefore, the remaining oil, 60%, is the target of the enhanced oil recovery called the tertiary oil recovery. There are several proven techniques for the tertiary oil recovery: surfactant-polymer flooding, foam flooding, CO_2 flooding, caustic solution flooding, microbial method, steam injection, and thermal combustion. Among the techniques mentioned above, the present paper focuses on the fundamental aspects of the surfactant-polymer flooding and foam flooding processes as well as related phenomena.

I. FORCES ACTING ON THE OIL DROP ENTRAPPED IN POROUS MEDIA AND THE ROLE OF SURFACTANT

At the end of the water flooding, the residual oil is in the form of oil ganglia trpped in the pore structure due to capillary forces as shown in Figure 1 (a). In a dynamic condition, the residual oil drop is mainly subjected to two forces: viscous forces of the displacing fluid and capillary forces. Viscous forces are known to be driving forces to mobilize the residual oil drop while capillary forces are known to be resistant forces to entrap the residual oil drop. Therefore, the trapped oil drop does not flow until the viscous forces of the displacing fluid exceed the capillary forces. It has been shown by several investigators that the capillary number, the ratio of viscous forces to capillary forces, is correlated with the level of the residual oil. An excellent review [1] summarizes various dimensionless forms of the capillary number. Even though the detailed variables of the capillary number are somewhat different depending on the investigators, the basic correlation between the capillary number and the level of the residual oil is identical. As the capillary number increases, the level of the residual oil decreases. A typical correlation reported by Foster [2] is shown in figure 2. When the capillary number is expressed by $N_{ca} = \mu V/\sigma\phi$ (where μ and V are the viscosity and velocity of the displacing fluid, σ is the interfacial tension and ϕ is the porosity), the value of the capillary number is around 10^{-6} after water flooding and should be increased by 3 to 4 orders of magnitude for tertiary oil recovery processes [3]. It can be easily shown from the definition of the capillary number that there are two possible approaches to increase the capillary number. One is to increase viscous forces, for example, by increasing viscosity and velocity of the displacing fluid. The other one is to decrease the interfacial tension. However, it is practically difficult to increase the

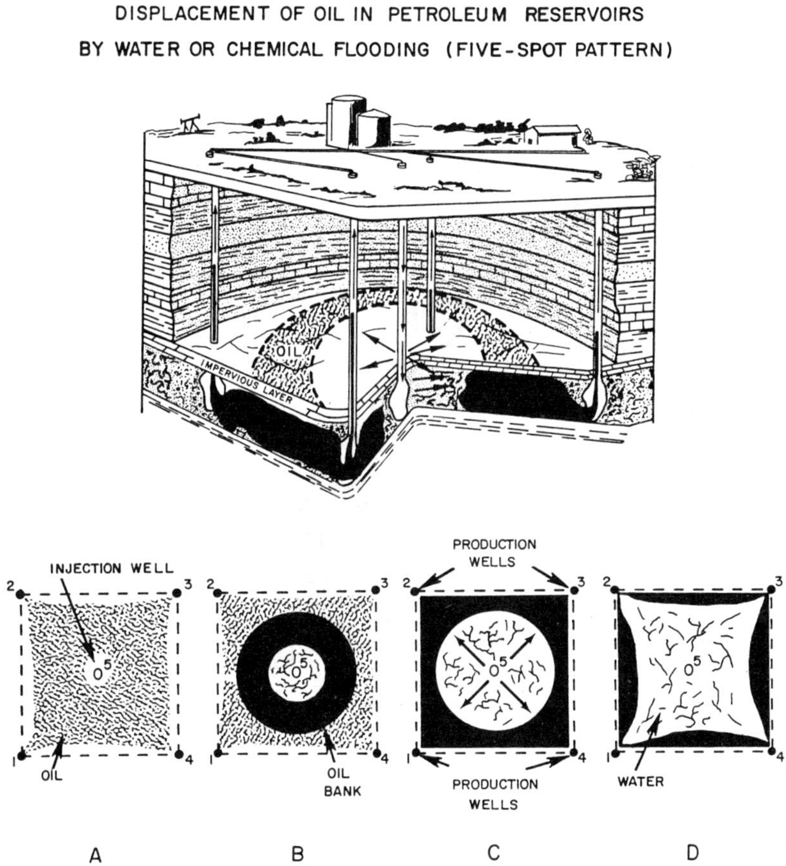

Figure 1. Schematic diagram of an oil reservoir and the displacement of oil by water or chemical flooding.

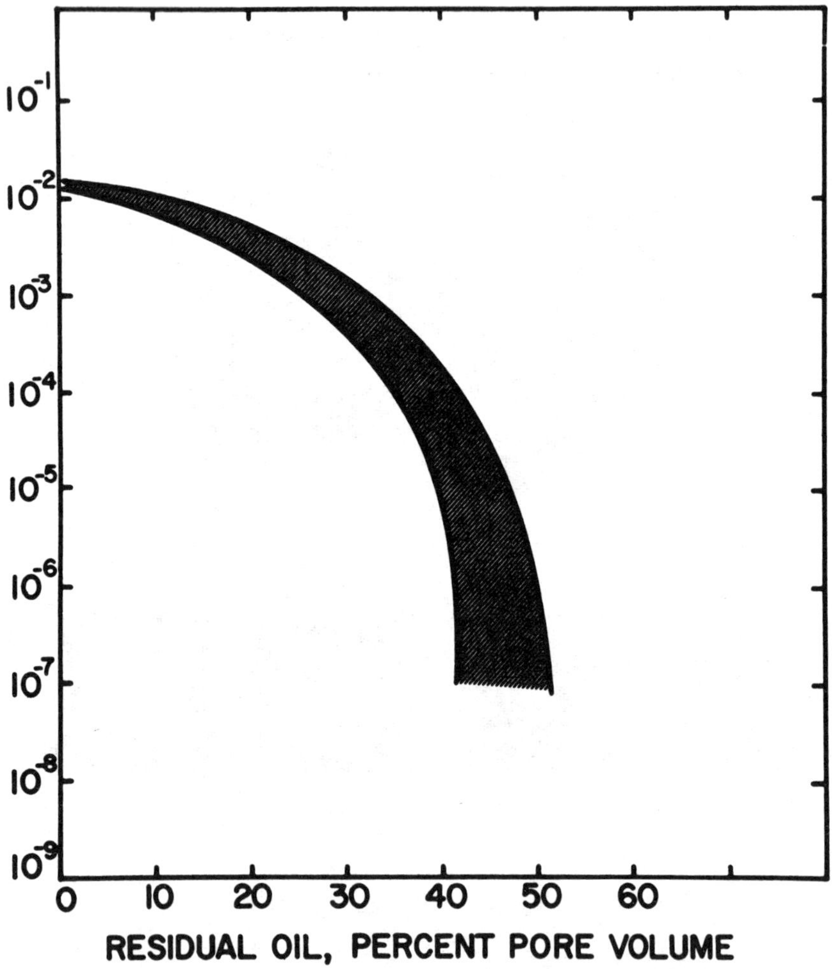

Figure 2. Dependence of residual oil saturation on capillary number (Foster, W.R., J. Pet. Tech., p. 206, Feb. 1973).

capillary number by 3 to 4 orders of magnitude using the former way. On the other hand, the use of appropriate surfactant can decrease interfacial tension from 20 or 30 dynes/cm to 10^{-3} or 10^{-4} dynes/cm. Therefore the surfactant-polymer flooding employs the use of surfactant to mobilize the oil entrapped in porous media by producing ultralow interfacial tension at the oil ganglia/surfactant solution interface. Figure 3 schematically shows the role of ultralow interfacial tension in promoting the mobilization of oil ganglia in porous media. In the surfactant-polymer flooding, surfactant is injected as the form of slug; the slug usually consists of surfactant, alcohol as cosurfactant, brine (or water) and sometimes oil.

A recent study [4] discussed the effect of capillary number on two-phase (oil-water) relative permeability curves. The study was done to determine whether the capillary number causes changes in the the two-phase permeabilities or whether one of its constituents, such as flow velocity, fluid viscosity, or interfacial tension, is the controlling variable. Two important results were: 1) that no velocity effects were noted within the range studied, and 2) that both the residual oil saturation and the imbibition-drainage histeresis decreased with an increase in the capillary number.

II. ROLE OF INTERFACIAL PROPERTIES OF OIL/BRINE INTERFACE

In the previous section, the role of surfactant is described as production of ultralow interfacial tension to mobilize oil ganglia in porous media. However, the mobilization of oil ganglia is not the only factor to achieve the good oil recovery. The mobilized oil ganglia should coalesce to form an oil bank [5]. A very low interfacial viscosity is desirable to form an oil bank [6,7]. Figure 4 shows the role of low interfacial viscosity in the surfactant-polymer flooding. The formed oil bank should be propagated through

Figure 3. Schematic diagram of the role of low interfacial tension in the surfactant-polymer flooding process.

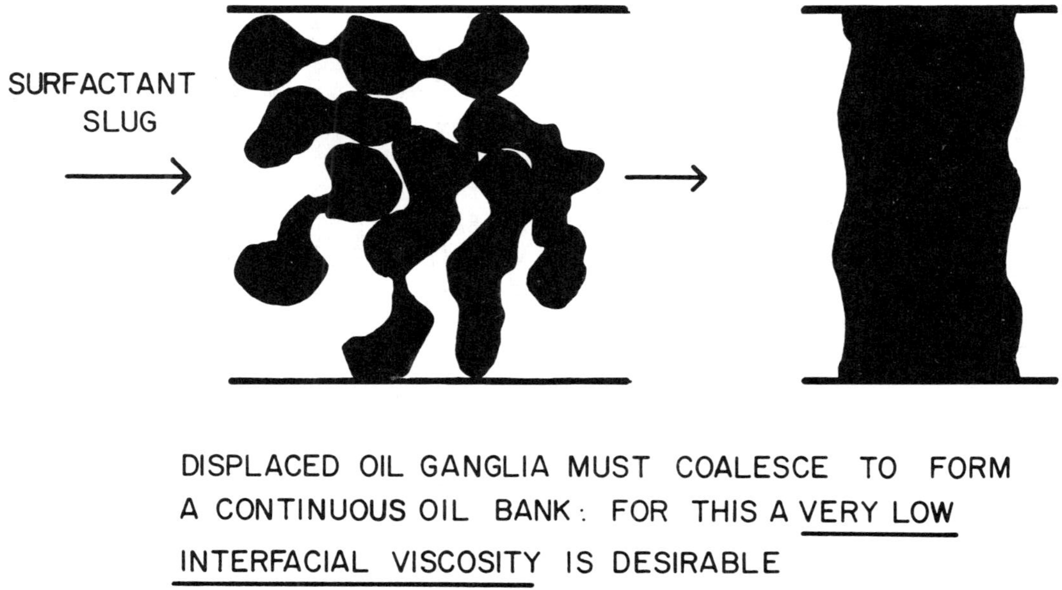

Figure 4. Schematic diagram of the role of low interfacial viscosity in the surfactant-polymer flooding process.

the porous medium without re-entrapment of the mobilized oil at the trailing edge of the oil bank. The maintenance of ultralow interfacial tension at the oil bank/surfactant slug interface is essential for minimizing the re-entrapment of the mobilized oil in the porous medium while the leading edge of the formed oil bank will coalesce with the oil ganglia. Since surfactants are relatively expensive chemicals, a small pore volume of the surfactant solution is used. Then the slug is displaced by a polymer solution followed by water or brine. Therefore, the mobility control of the different fluids is another essential factor for the good oil recovery. The mobility control is basically achieved by increasing the viscosities of the fluids in the order of injections. Only exception occurs in the case of water after an injection of a polymer solution due to the economic consideration. The studies on the mobility control can be found in the previous papers [8-10]. The Figure 5 schematically illustrates the movement of the oil bank, surfactant slug and the polymer solution in the porous medium. Other important factors to be considered for the good oil recovery are: 1) the mass transfer of surfactant to the oil bank 2) the wettability of rock 3) surface charge at the oil/brine as well as rock/brine interfaces 4) and the interaction between surfactant and polymer. In the following section we briefly review these factors: 1) Chiang and Shah [11] reported that not only the equilibrium values of the parameters such as interfacial tension and interfacial viscosity are important for oil recovery but the dynamic process of surfactant partitioning is also important in mobilization of oil ganglia. The higher the efficiency of mass transfer of the surfactant to the oil bank, the easier the deformation and mobilization of oil ganglia. 2) The choice of surfactant can influence the wettability of the rock surface to oil and brine [1] which in turn influences the coalesence of oil ganglia. 3) Chiang and Chan [11,12] observed that interfacial charge density at the oil/brine interface is an important factor which can affect interfacial tension and interfacial viscosity at the oil/brine interface as well as oil recovery efficiency. 4) Trushenski [13] has shown that surfactant-polymer

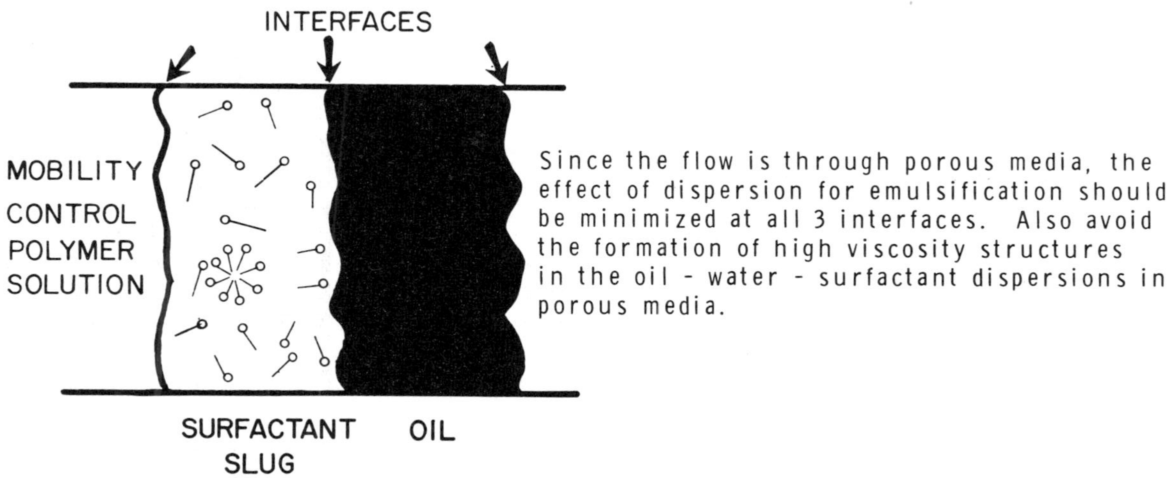

Figure 5. Schematic illustration of the effects of dispersion and emulsification between the various slugs during the surfactant-polymer flooding process.

incompatibility leading to a phase separation of surfactant and polymer strikingly reduces the efficiency of the process.

III. SURFACTANT-POLYMER FLOODING

A. CLASSIFICATION OF THE SURFACTANT-POLYMER FLOODING

The surfactant-polymer flooding can be largely divided into two concepts based on the concentration of surfactant in the slug [14]. One concept uses a large pore volume of a low concentration surfactant solution. The other uses a small pore volume of a high concentration surfactant solution. The present paper will focus on the latter concept, a high concentration surfactant-polymer flooding. The high concentration surfactant slug can be further divided into an aqueous surfactant slug and a soluble oil slug depending on whether the surfactant is dissolved in aqueous phase or in oil phase, respectively [15, 16].

B. THE CONCEPT OF THE OPTIMAL SALINITY

A microemulsion is formed when oil and water (or brine) are mixed with suitable surfactant and alcohol. The structure of microemulsion varies depending on the given conditions and composition of the system. The typical way to see the phase behavior is to increase the concentration of salt while other parameters such as concentrations of given surfactant and alcohol, temperature, type of oil, etc. are fixed. At low salt concentrations, surfactant is in the brine phase and solubilizes oil to form an oil-in-water microemulsion existing as lower phase. As the concentration of salt increases, surfactant starts to form middle phase. Several authors have proposed various structures for the middle phase microemulsion, e.g., bicontinuous structure and the coexistence of oil-in-water and water-in-oil microemulsions [17-24]. At high salt concentrations, surfactant is in oil phase to form water-in-oil microemulsion.

The mechanism of the formation of middle phase microemulsion can be described as follows [7]: as one increases the salinity, the CMC decreases, the aggregation number of the micelles increases and the solubilization of oil within micelles increases. The compression of the electrical double layer around micelles will occur, hence reducing the repulsive forces between the micelles . Thus the reduction in the repulsive forces and increase in the attractive forces between the micelles will bring the micelles closer and ultimately lead to a separation of a micelle rich phase forming the middle phase microemulsions.

Figure 6 shows the parameters leading to the transition from lower phase to middle phase to upper phase microemulsion. Thus, by choice of a suitable parameter, one can obtain the transition in the structure of these microemulsions. At optimal salinity, the partition coefficient of surfactant in the excess oil and brine phases is found to be near unity and the interfacial tension between the excess oil and excess brine is also minimum [12]. The significance of the optimal salinity concept for enhanced oil recovery is shown in Figure 7. It is evident that the oil recovery is maximum at optimal salinity for the system reported here. This phenomenon has been reported in the other study [25]. An excellent correlation between the capillary number and oil recovery is also evident from Figure 7 [26]. Hsieh [27] observed that as the oil chain length increases, the optimal salinity increases and the volume of the middle phase decreases. The range over which the middle phase microemulsion exists also increases as the oil chain length increases.

A recent study [28] defined the contact angle for optimal salinity which was related to the previously introduced optimal salinity in particular to that associated with best oil recovery. The interesting result was that the optimum attainable contact angles for microemulsion-based oil recovery was not zero.

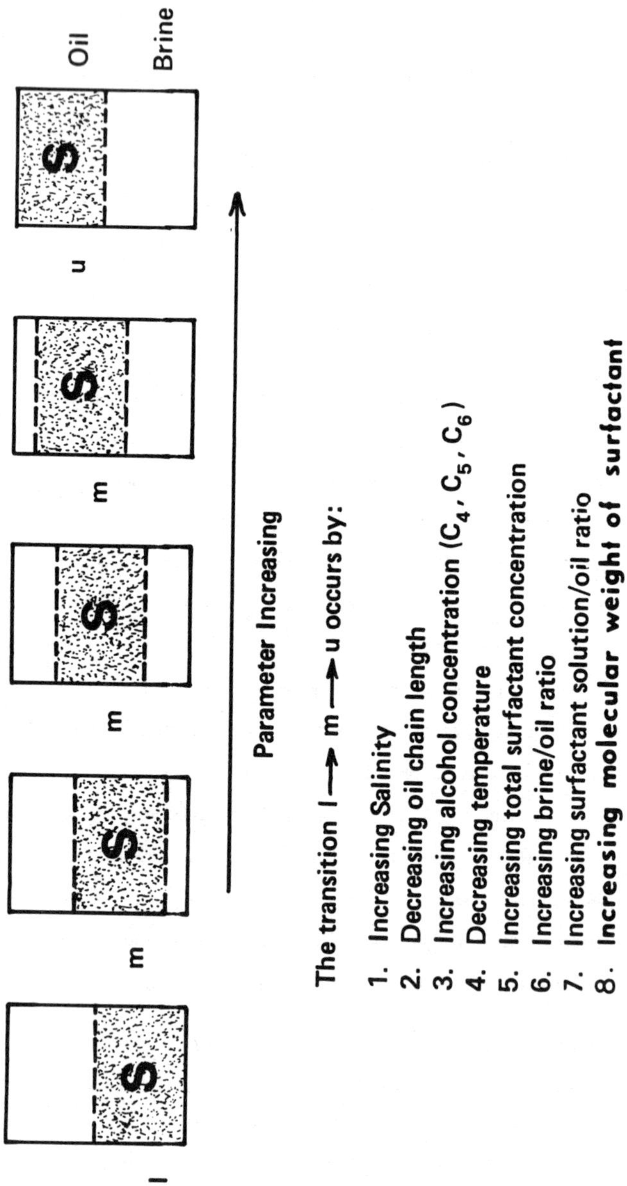

Figure 6. Schematic illustration of the factors influencing the L → M → U transition in surfactant/oil/brine/alcohol systems.

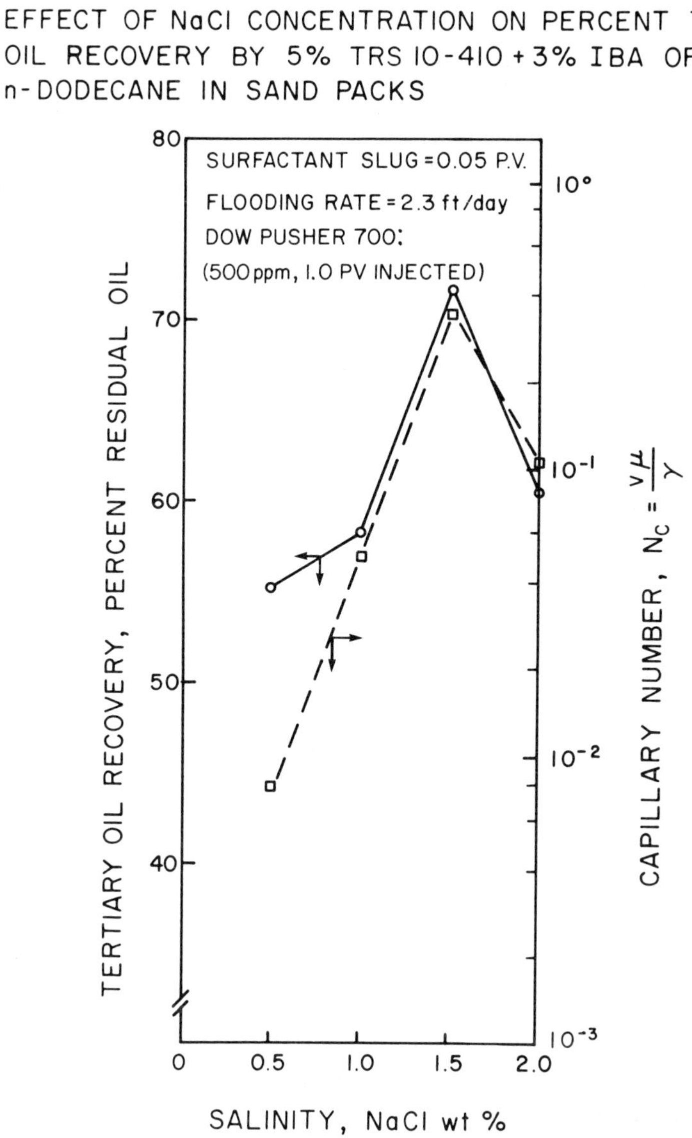

Figure 7. Effect of NaCl concentration on tertiary oil recovery in sand packs.

C. COALESCENCE PROCESS

The oil recovery process is not a static process but a dynamic process. Several different phases meet at the interfaces resulting in the existence of mixing zones. Oil drops form and coalesce during their flow through porous media. The previous study [5] showed that tertiary oil recovery efficiency correlated almost directly with coalescence rate for the systems studied; the oil produced by a rapidly coalescing system is primarily emulsion-free. However, the oil produced by a slowly coalescing system can be partially emulsified. In this case, the final oil recovery was not maximized because the slowly coalescing system left more re-entrapped oil in the porous medium.

Vijayan and co-workers [29] have shown that minimal separation time or the fastest coalesence rate of emulsions occurs at the optimal salinity. Also, they reported that the pressure drop across a porous medium was minimum at and around the optimal salinity when emulsions prepared at various salinities were injected in the porous medium.

Figure 8 shows a correlation between the coalescence rate in emulsions and the apparent viscosity in the flow through porous media. The minimum apparent viscosity for the flow of emulsions in porous media coincides with minimum phase separation at the optimal salinity. This correlation between the phenomena occuring in porous media and outside the porous medium allows us to use coalescence measurements as a screening criterion for many surfactant formulations for their possible behavior in porous media. Figure 9 summarizes all the phenomena occuring at the optimal salinity in relation to enhanced oil recovery by surfactant-polymer flooding.

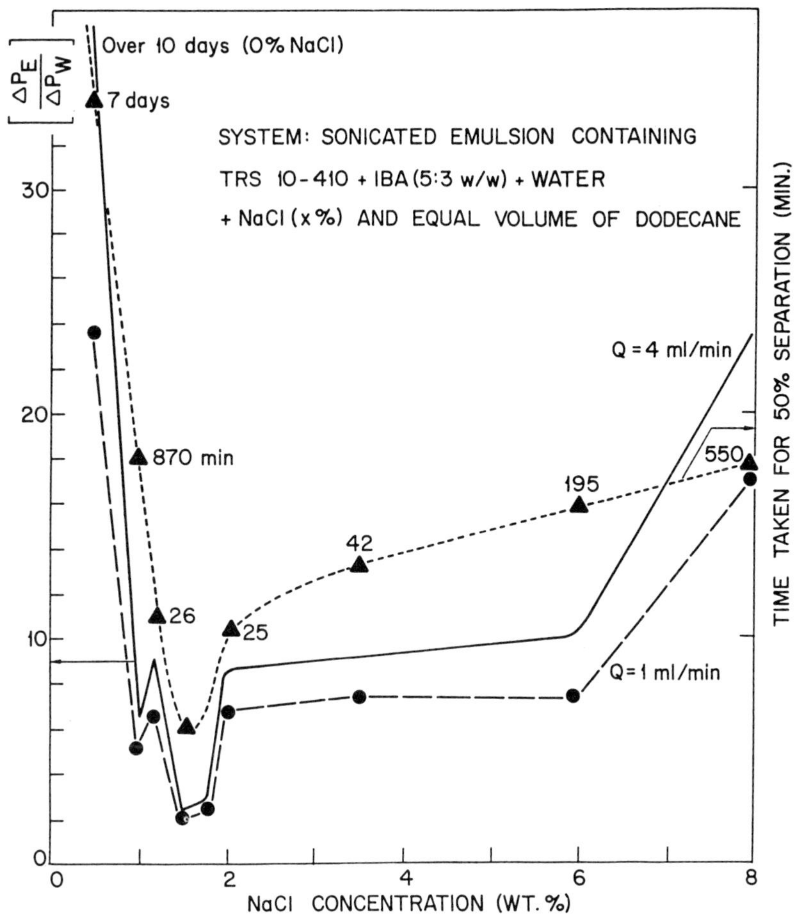

RATIO OF PRESSURE DROP OF EMULSION TO WATER IN A SAND PACKED COLUMN (1/4" dia. & 2" long) WITH NaCl CONCENTRATION, AND TIME TAKEN FOR 50% SEPARATION OF EMULSIONS

Figure 8. A correlation between the apparent viscosity and coalescence rate of sonicated emulsions.

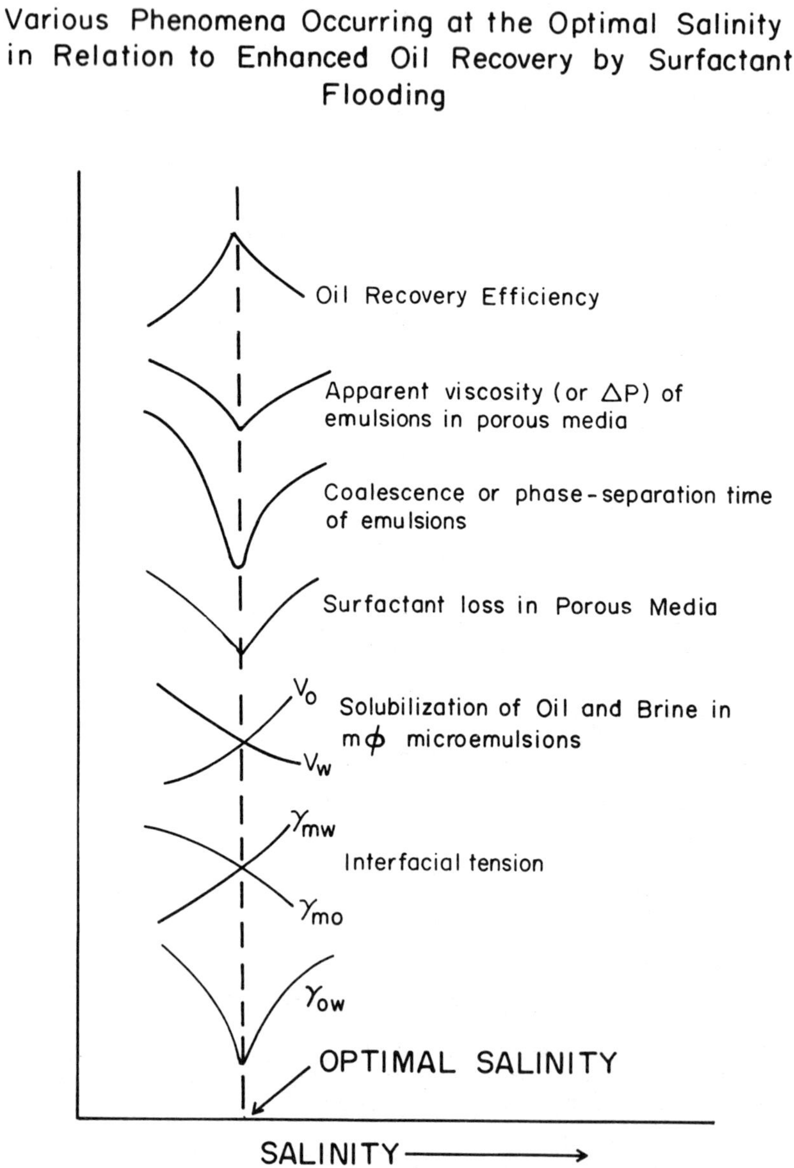

Figure 9. A summary of various phenomena occurring at the optimal salinity in relation to enhanced oil recovery by surfactant-polymer flooding.

D. EFFECT OF ALCOHOL

In general, a surfactant formulation for enhanced oil recovery includes a short chain alcohol (e.g. C_4 to C_6 alcohols and their isomers). The possible effects of alcohol include the change in viscosity, lowering of the interfacial tension, reduction in interfacial viscosity or change in surfactant partitioning and modifying the solubility of surfactant in oil or brine phase. A previous study [11] suggested that the alcohol promotes the mass transfer to the interface and a rapid reduction in the magnitude of the interfacial tension.

A recent study [30] showed that the separation of alcohol from surfactant during the flow through a porous medium leaded to poor oil recovery. Even though the change in the composition of the surfactant slug cannot be avoided during the surfactant-polymer flooding due to the dilution and chromatographic effect of a porous medium, the separation of alcohol from surfactant is minimized at the optimal salinity. The peaks of recovered surfactant and alcohol in the produced fluids during the laboratory experiments coincided at the optimal salinity while they were separated at the non-optimal salinity [30].

Lam [31] did a microvisual study of residual oil mobilization that used a relatively simple chemical system. It was found that under certain circumstances the capillary number required to mobilize the oil drop could be reduced greatly if the displacing phase contained more alcohol than it would be in equilibrium with the trapped phase.

E. SOLUBLE OIL SLUG

In general, the surfactant slug in the surfactant-polymer flooding is an aqueous solution containing surfactant and alcohol. However, a soluble oil is a water-in-oil microemulsion; the continuous phase is oil.

Since soluble oil is usually compatible and miscible with the residual oil, the residual oil easily coalesces and forms the oil bank upon injection of soluble oil; the interfacial tension between oil and the slug is almost zero. The trailing edge of the slug is diluted by aqueous drive solution and may be transformed from water-in-oil microemulsion to oil-in-water microemulsion, which can be easily displaced by aqueous drive (polymer) solution. However, the interfacial tension at the trailing edge of the slug should be low enough to avoid the entrapment of the slug. The breakthrough of surfactant is delayed until the oil comes out because the surfactant in soluble oil is preferentially soluble in oil. The oil recovery mechanism is well described in the previous study [32]. The previous study [16] described the screening method of soluble oils by defining a viscosity-solubilization index (VSI); a low viscosity and a high solubilization are required for a good soluble oil.

A recent study [33] indicated that a soluble oil formulated by anionic surfactant with nonionic surfactant could be utilized for highly viscous oil recovery. These researchers formulated the soluble oil using Aerosol-OT, Arlacel 20, mineral oil, and water. The study indicated as shown in Figure 10 that the solubilization of water in the soluble oil was related to the oil recovery; the oil recovery was maximum with the formulation which showed the maximum solubilization of water in the microemulsion. The study also showed that the soluble oil which solubilized more water displaced the residual oil in a piston-like displacement as shown in Figure 11. When the surfactant formulation exhibiting the highest solubilization of water was used, oil bank was nicely formed and maintained while oil bank was invaded by water at the early stage of displacement when the surfactant formulation showing poor solubilization of water was used.

One drawback in using soluble oil is that the solubilization of

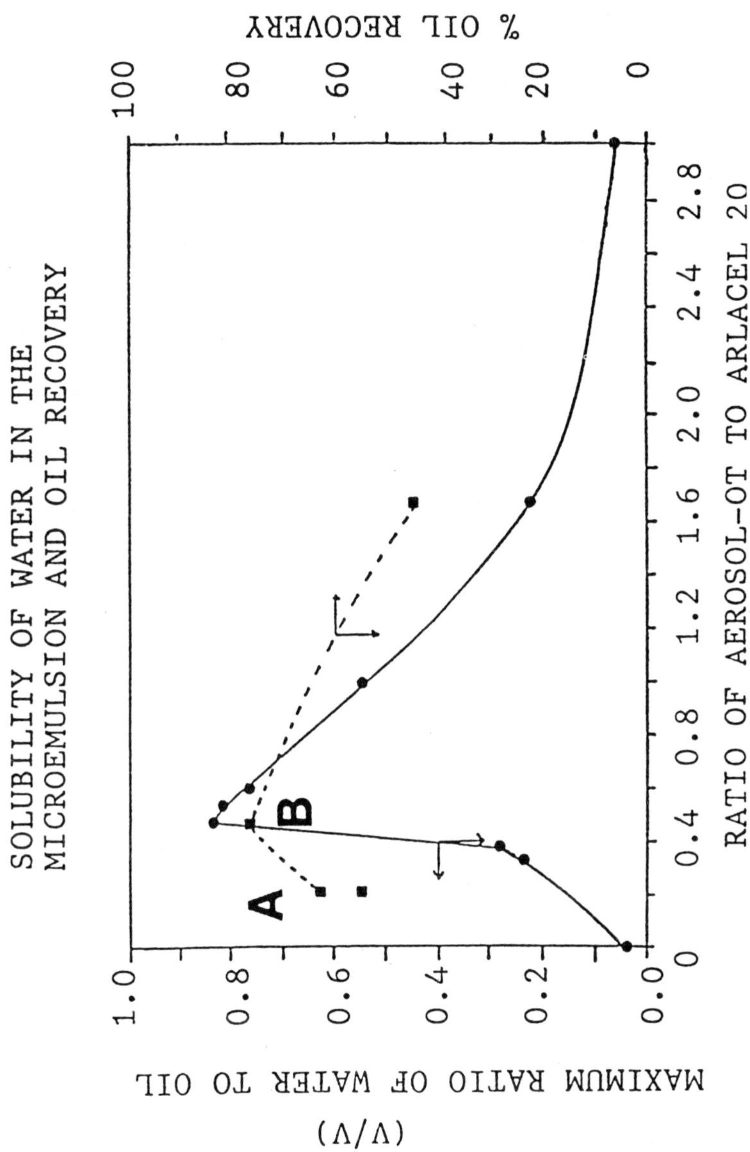

Figure 10. A correlation between the solubility of water in microemulsion and the oil recovery.

OIL PRODUCTION HISTOGRAM

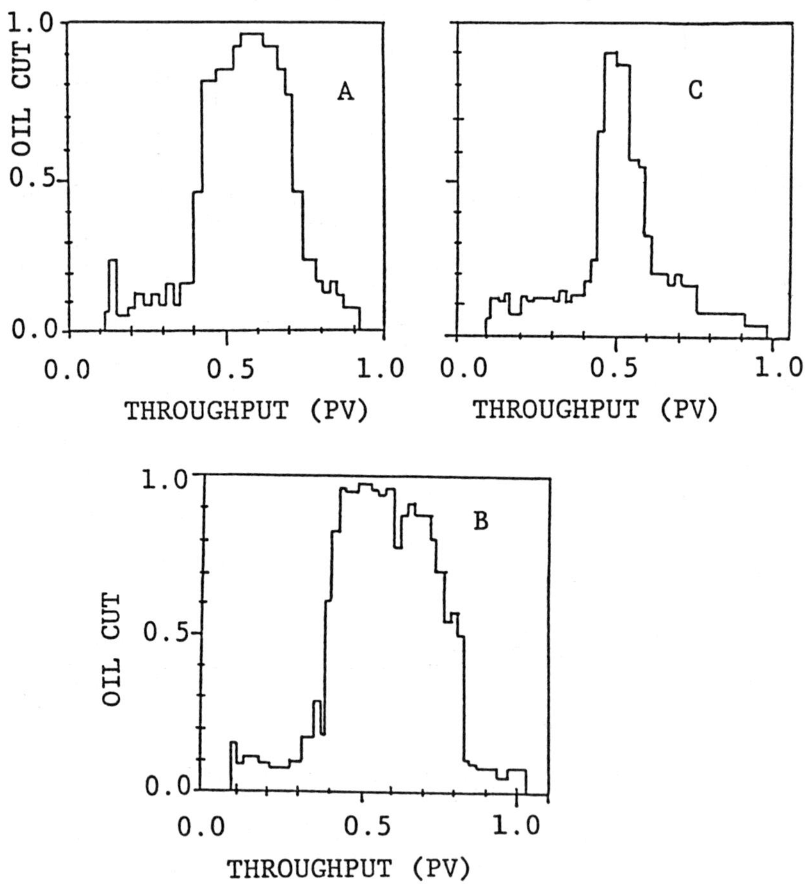

Figure 11. Oil production histograms of the alcohol-free water-in-oil microemulsion flooding for heavy oil recovery.

water decreases as the salt concentration increases. Therefore, the fresh water is usually used to prepare the soluble oil and the fresh water is also used after the soluble oil slug [34].

IV. THE SURFACTANT-POLYMER FLOODING FOR HIGHLY VISCOUS OIL

As mentioned previously, a nonionic surfactant with an anionic surfactant showed the possiblility that the surfactant-polymer flooding could be used for highly viscous oil. The main reason that the surfactant-polymer flooding has not been used for highly viscous oil is that the solubilization of oil in microemulsion decreases as the chain length of oil increases resulting in high interfacial tension [34]. However, a recent study [35] showed that the volume of middle phase formed using a nonionic surfactant with an anionic surfactant was much larger than the volume of middle phase formed using a short chain (C_4-C_6) alcohol with anionic surfactant. Oil recovery experiment using a sand pack also showed that the middle phase microemulsion containing nonionic surfactant recovered much more oil than the middle phase microemulsion containing alcohol.

A previous study [34] also showed that, when light hydrocarbon such as benzene, toluene or cyclohexane is added to the active components, the microemulsion became very effective for bituem recovery from sands.

V. FOAM FLOODING

Foams are aggregates of dispersed gas bubbles in a relatively small amount of liquid. Bubbles separated from each other by thin liquid films vary in size from several microns to several mm. Foams are both unusual and intriguing in their physical properties, and have been studied by many researchers [36-44]. There is little doubt that at least some foams behave like non-Newtonian fluids, and have apparent viscosities considerably higher than those of either gas or liquid phase [45]. The foam properties are determined by numerous factors such as surface tension of the liquid, bubble size distribution, interbubble gas diffusion, surface viscosity, Gibbs elasticity, drainage of liquid from lamellae, adsorption/desorption of solute at liquid-gas interface, the rheology of the adsorbed layer, external pressure, and temperature. For the flow of foam through a porous medium, the permeability of porous medium, pore size, and surface properties of matrix etc. have also to be considered as the factors related to rheological properties of foams.

Foam has been successfully used as a fracturing fluid for several years [46]. It is a very powerful tool for fluid leakoff control [47,48]. Montman [49] has reported that foam cement could be used to seal underground storage caverns, insulate wellbores, and perform remedial squeeze jobs. Recently, field report [50] showed that foam cement can develope adequate compressive strength and keep cement slurry density below the hydrostatic breakdown gradient of the sensitive formations, therefore a fallback problem can be avoided. In the meantime, CO_2 foam has been employed in gas and oil wells, high-and low-temperature reservoirs, and deep and shallow holes even though there are some limitations. A successful field application of CO_2-foam fracturing fluids in the Arkansas-Louisiana-Texas region was reported by Warnock et al. [51].

One technique for thermal oil recovery involves the injection of steam into a reservoir to reduce oil viscosity and make the oil more mobile. Foam has been suggested as a "blocking agent" in steam injection to reduce gravity override and channeling effects (figure 12). The mobility of injected steam can be controlled by the generation of foam in porous medium and the reduction of mobility is proportionately higher in the more permeable sands. The process is highly efficient since the foam first finds its way into the largest pores, in which it tends to block further flow. The smaller pores are thus invaded next, and so on until the entire permeable section has been filled

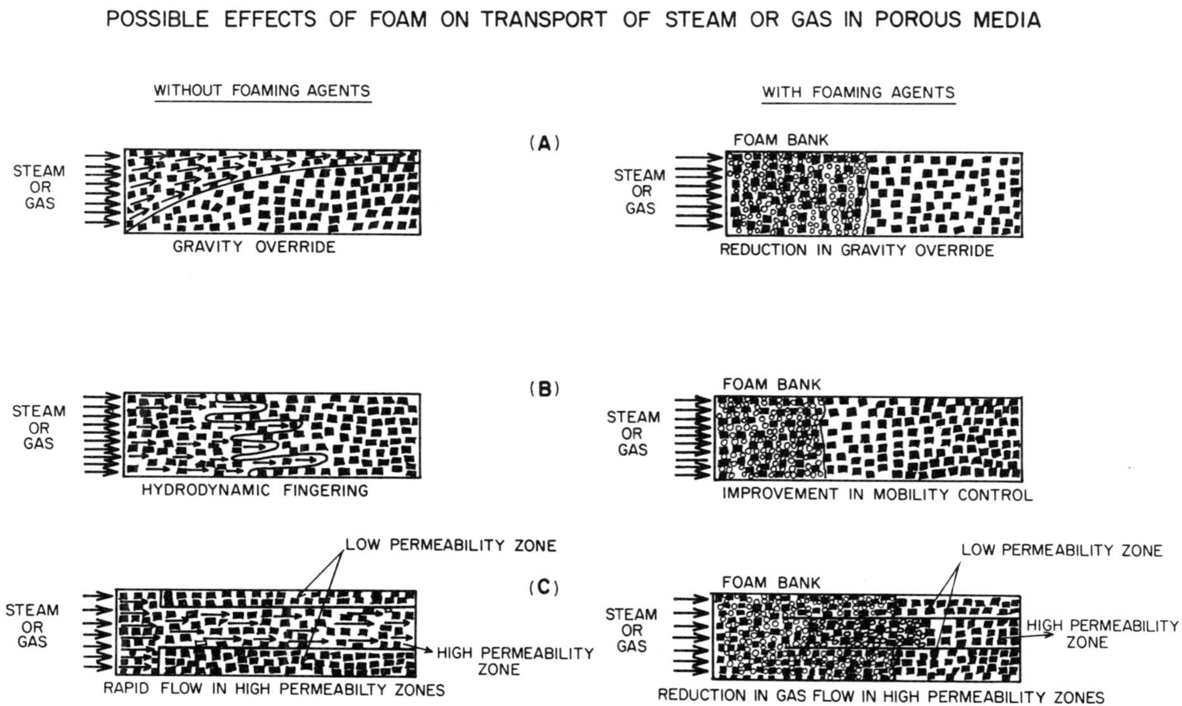

Figure 12. Schematic illustration of the effects of foam on transport of steam or gas in porous media.

with the foam. Laboratory experiments have shown that vertical sweep efficiency is much improved by the injection of foam [52]. Despite the great potential of the foam flooding process, the mechanisms of foam/oil displacement are still not fully understood. The basic principles of foam flooding have been studied for the past 25 years by many investigators [53-57]. A number of research papers have been published on foam behavior in porous media [58-62].

This section presents our results of the laboratory studies designed to correlate the relationships between the surface properties and microscopic characteristics of foaming agents and fluid displacement in porous media. Pure as well as commercial surfactants were selected and the concept of chain length compatibility was applied to both systems. The observations of microscopic foam behavior in porous micromodel were also reported. A novel concept, namely, surfactant-polymer-foam flooding, is proposed for heavy oil recovery.

A. INTERFACIAL PROPERTIES OF FOAMING AGENTS

a. Surface Tension

For pure surfactants such as sodium dodecyl sulfate (SDS), it was observed [63] that the surface tension was the lowest for system consisting of SDS and dodecyl alcohol as compared to that of SDS with other long chain alcohols. The situation becomes complex for the mixed surfactant systems (SDS + $C_nH_{2n+1}OH$) in the presence of brine since the precipitation of mixed surfactants occurs. Sharma and Shah [64] have shown that the addition of NaCl to the foaming system of SDS + $C_{12}OH$ results in an increase in surface tension. This increase in surface tension is presumably related to the decrease in solubility of dodecyl alcohol in the solution and the precipitation of the complex of SDS+$C_{12}OH$. For commercial surfactants such as Suntech IV and Stepanflo 40, the surface tensions are also influenced by the addition of long chain alcohols. Interestingly, the lowest surface tension was found for both foaming solutions with the addition of dodecyl alcohol. The concentrations of the mixed surfactant system for above observations were above the critical micelle concentration.

A decrease in surface tension was observed [64] with the increase of temperature for both pure and commercial foaming agents without adding electrolytes or alkyl alcohols. As the temperature is increased, the kinetic energy of molecules increases, and the attractive forces between molecules are easily overcome, which in turn reduce the surface tension of the foaming solutions. If the effect of electrolytes and alkyl alcohol are considered, the situation becomes much more complicated. Basically, a similar trend in the variation of surface tension with temperature was observed [65] for both foaming solutions in the absence or presence of brine.

b. Foaminess

In the pre-CMC region, foaminess generally increases with the increase of surfactant concentration. The maximum foam volume was observed when the concentration was around or slightly above CMC. Further increase in concentration does not improve foaminess. Usually, the lower the CMC, the more efficient the foaminess of surfactants. But the relationship is not always true. Sharma and Shah [64] reported that the foaminess was maximum when both the components of the surfactant system had the same chain length. Shah and Shiao [66] also have shown that the surfactant molecules pack tightly in the mixed film when the chain length of both components is equal.

The foaminess of both pure and commercial surfactants increased with and increase in temperature [64]. At room temperature, a decrease in foaminess was observed for the mixed surfactant systems with the addition of NaCl presumably due to the precipitation of mixed surfactants or the decrease of solubilization of alkyl alcohols in the solution. On the other hand, the effectiveness of brine to reduce foaminess decreased at higher temperatures. Since so many factors affect the foaminess, there is still no clear relationship found between foaminess and surface properties of foaming agents.

c. Foam Quality

The volumetric gas content of a foam is called foam quality and is expressed by the following relationship:

foam quality = (gaseous volume)/(total foam volume)

Foams with a quality higher than 0.8 are termed "dry" while those with a quality lower than 0.7 are termed "wet". Raza and Marsden [40] have shown that a minimum of about 4% liquid is required to produce the foams. It implies that the maximum foam quality which can be achieved is 0.96. Lemlich [67] reported that the volume fraction of liquid in the foam can be determined by measuring the specific conductivity of foam $\phi_i = 3K_f/K_s$

where ϕ_i = the volume fraction of liquid in the foam

K_f = the specific conductivity of foam

K_s = the specific conductivity of solution

The foam quality is equal to $1 - \phi_i$. The principle of foam conductivity is based on the view of foam as a network of interconnected films whose liquid content has the same specific conductance as the bulk solution.

The foam quality of mixed surfactants as a function of alkyl chain length at different time intervals (5 min., 10 min., and 20 min.) is shown in figure 13 [68]. Foam quality varies with time because the static foam collapses gradually and the liquid is drained from the thin liquid film.

For the mixture of commercial surfactant Stepanflo 40 with different long chain alcohols, foams were considerably "dry" (0.84 - 0.92). The foam quality decreased with the increase of hydrocarbon chain length of alcohols from C_8 to C_{10}. There was almost the same effect on foam quality by adding decyl or dodocyl alcohol in Stepanflo 40 solution. Interestingly, the foams generated by adding hexadecyl alcohol was even "dryer" than that without adding any alcohol.

d. Surface Viscosity

Surface viscosity is a measurement of molecular interactions in monomolecular films at the interface. The study of surface viscosity is useful in connection with the stability of foams. A high surface viscosity leads to a retardation of bulk liquid flow near the surfaces, which in turn reduces the rate of thinning of the liquid films [69] and, consequently increases the foam stability.

Shah and co-workers [66,70] have shown that the increase in foam stability of decanoic acid solution in the presence of decanol is due to the increase in surface viscosity caused by the alcohol. It was observed [63] that the surface viscosity was the highest when both components had the same chain length (i.e., SDS + $C_{12}OH$). These results suggest that the surfactant molecules having similar chain length are tightly packed at the interface as compared to the

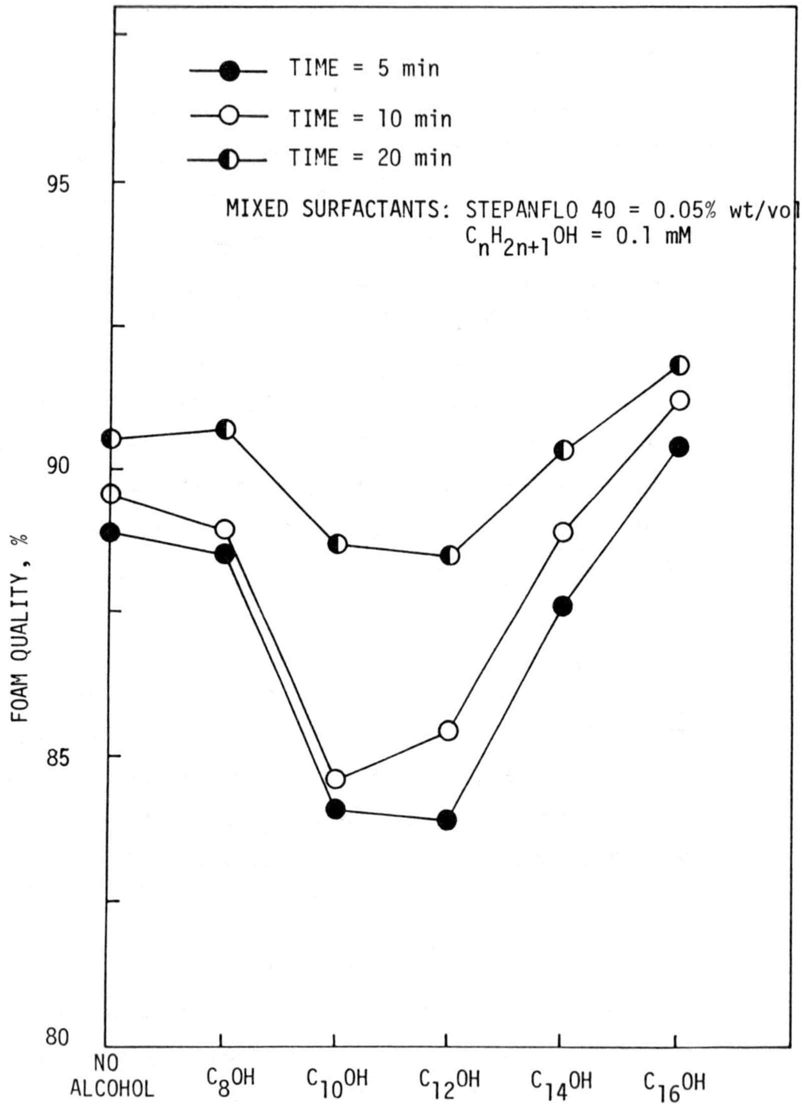

Figure 13. VARIATION IN FOAM QUALITY AS A FUNCTION OF STEPANFLO 40 WITH DIFFERENT ALKYL ALCOHOLS

molecules with the dissimilar chain length.

e. Bubble Size Distribution

Figure 14 shows the bubble-size distributions of various foams at 15 minutes interval after the foams were generated [63]. The bubble size increases with elapsed time due to coalescence. The average bubble size was a minimum for the mixed surfactants having the same chain length (i.e., SDS + $C_{12}OH$) and increased with increasing difference in the chain length. Sharma et al. [65] also reported that the bubble size decreases with increasing temperature.

The rate of foam decay resulting from interbubble gas diffusion and gravity drainage has been shown [71] theoretically and experimentally to be a sensitive function of the initial distribution of bubble sizes. This may explain why most of foam stability tests are difficult to reproduce. The rate of gas diffusion between bubbles and the mechanism of gravity drainage can be dramatically changed by surfactant molecules' packing pattern at the air/liquid interface which in turn affects the collapse of foams. However, based on relative comparison, the measurements of bubble size distribution for the mixed surfactant solutions are still very useful in predicting the foam stability.

B. MICROSCOPIC VIEW OF FOAM FLOW IN POROUS MEDIA

Figure 15 shows the photomicrographs of foams generated inside the micromodel [72]. It is evident that the mixed surfactants of same chain length (SDS+$C_{12}OH$) produced smaller foam bubbles and large number of lamellae as compared to the surfactant system with dissimilar chain length. These photomicrographs indicate that the injection of a gas in the micromodel filled with surfactant solution can generate foam in-situ.

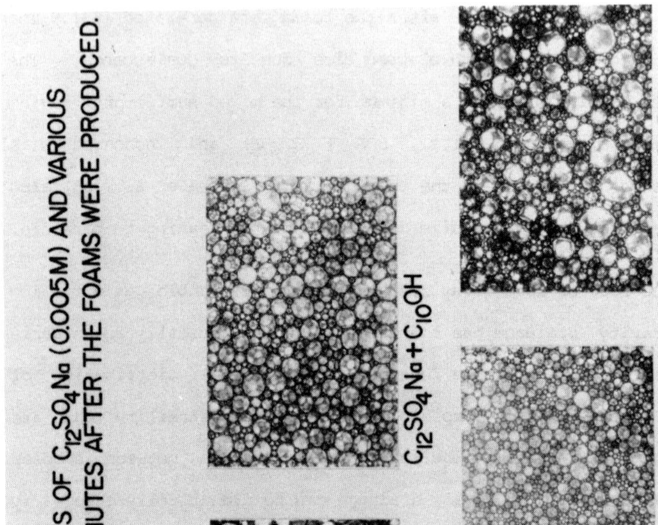

Figure 14. Bubble size for foams of mixtures of SDS (5 mM) and various alkyl alcohols (0.5 mM) at 15 minutes after the foams were produced.

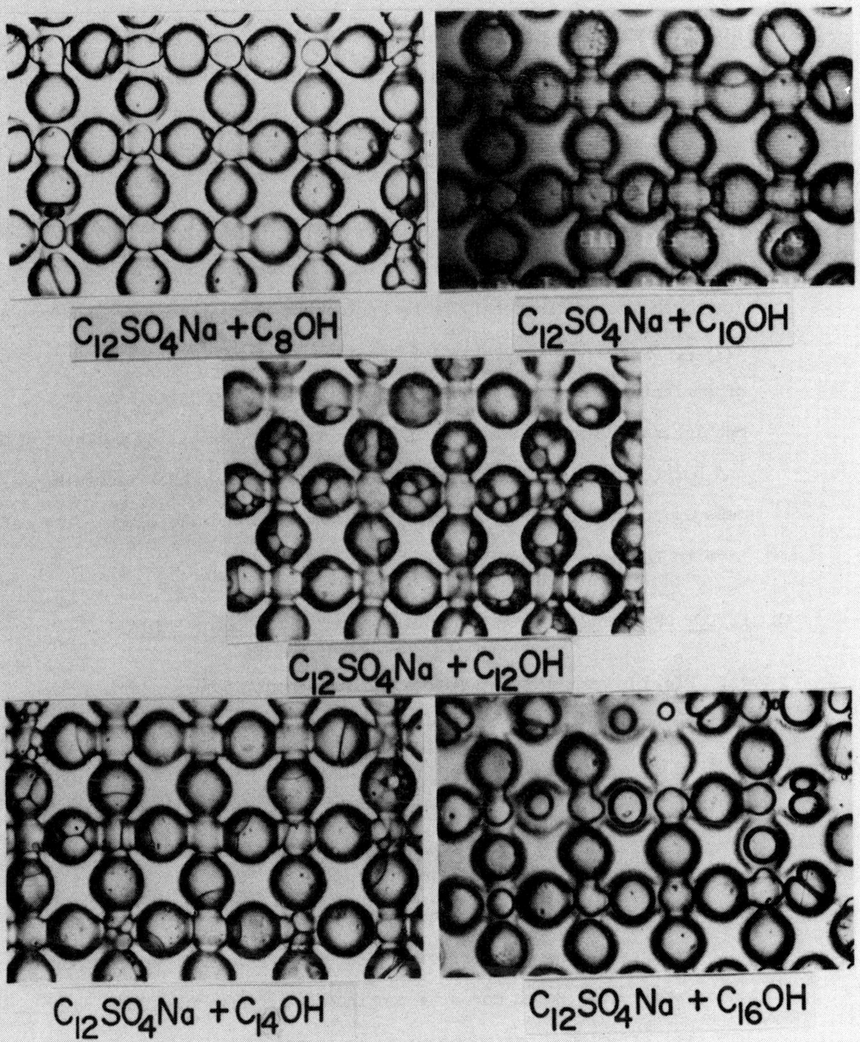

Figure 15. Photomicrographs of foams in micromodel (pore neck 50μm x pore diameter 100μm).

Many researchers [73-76] have studied the rheology of foam in porous media. They pointed out that the flow behavior of foam in a porous media is a complex one, which cannot be correctly described in terms of the high apparent viscosity or bubble stability of foam. There is a general agreement that foam behaves like a pseudoplastic fluid with high apparent viscosity. However, there are diverse opinions on how foam and its components are transported through the network of pores.

The mechanism of foam flow through porous media described by Holm [75] is of a moving continuous liquid phase and a discontinuous, internal gas phase moving discretely or by rupture of the gas cells. Shah et al. [77] proposed two mechanisms of gas flow in porous media in the presence of surfactant solution (figure 16). Foams with high stability follow the bubble-train mechanism, whereas, foams with low stability follow the pop and burst mechanism in a porous medium. Of course, this mechanism also depends on the amount of foam generated in-situ and the bubble size relative to the pore size.

C. EFFECT OF SURFACTANTS ON FLUID DISPLACEMENT IN POROUS MEDIA

a. Fluid Displacement in Porous Media without Foaming Agents

The variation in fluid displacement efficiency and breakthrough time as a function of applied air pressure (ΔP) in sandpacks and Berea cores are shown Figure 17 and 18, respectively [65]. The fluid recovered from sandpacks or Berea cores was very low because of channeling effect and gravity override of air. For Sandpacks, fluid displacement efficiency slightly decreased as ΔP increased. However, for Berea cores, a continuous increase in fluid displacement efficiency was observed up to 25 psi and remained constant with

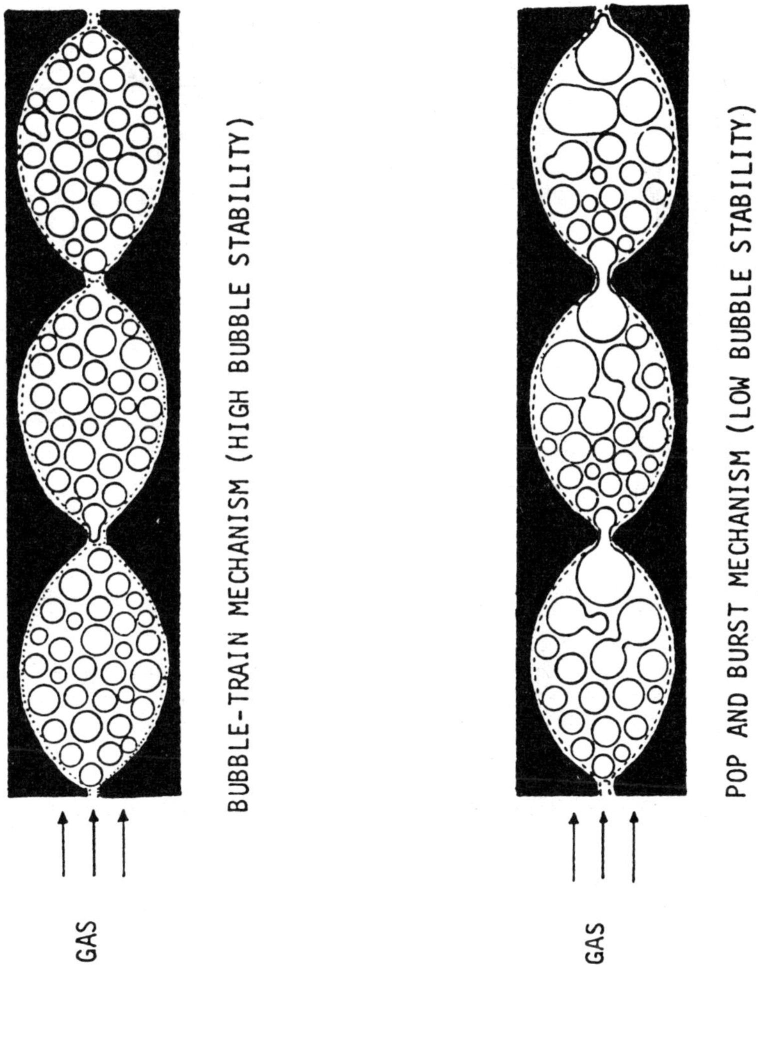

Figure 16. Two mechanisms of gas flow in porous media filled with surfactant solution.

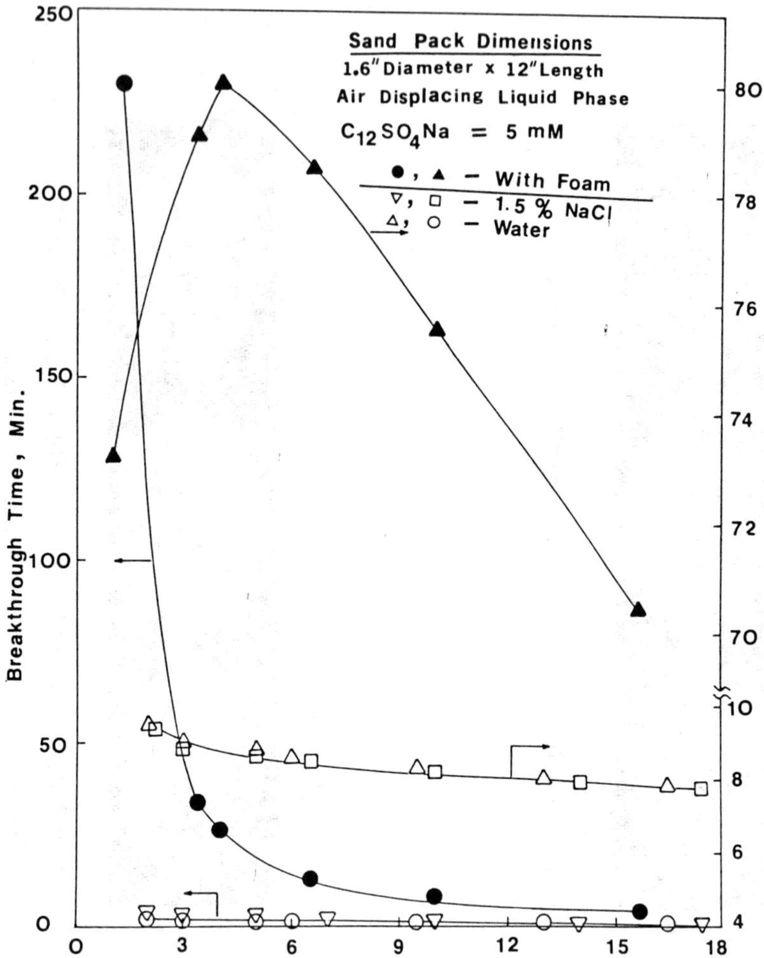

Figure 17. Effect of ΔP on breakthrough time and fluid displacement efficiency in sandpacks with and without foaming agents.

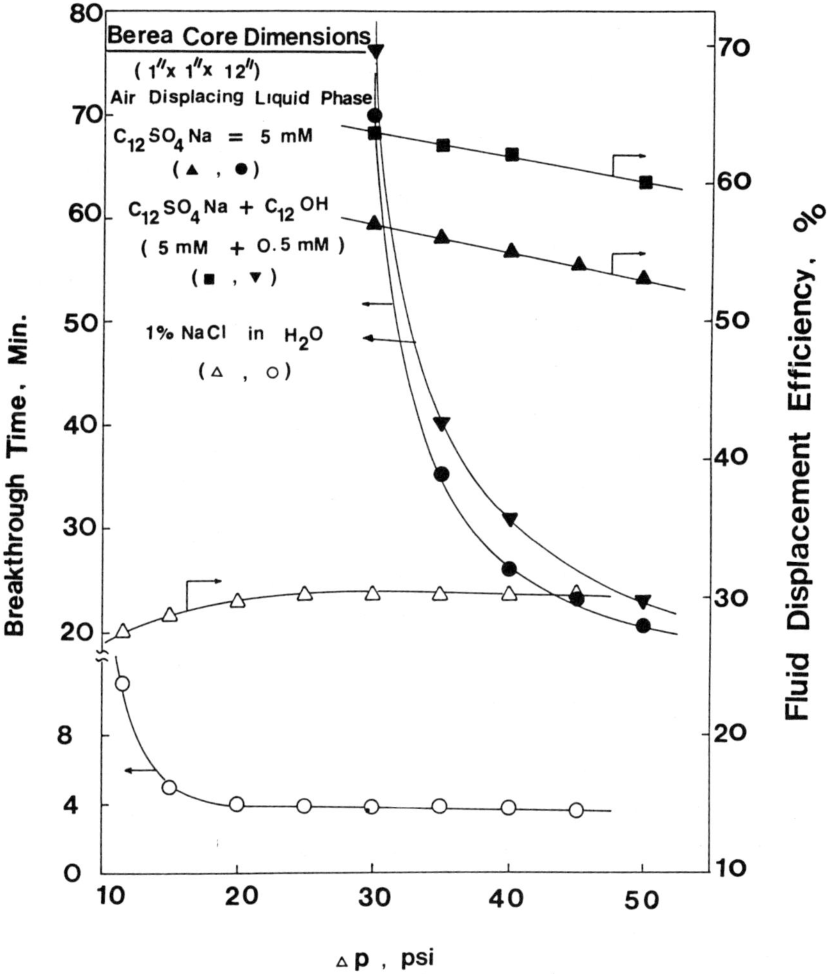

Figure 18. Effect of ΔP on breakthrough time and fluid displacement efficiency in Berea cores with and without foaming agents.

further increase in ΔP. A sharp decrease in breakthrough time was observed for both cases at low ΔP and remained constant at high ΔP. It is clear that the air mobility has to be controlled to reduce the channeling effect and gravity override which in turn improve the fluid displacement efficiency.

b. Fluid Displacement in Porous Media with Foaming Agents

It is well known that both the fluid displacement efficiency and the breakthrough time are increased in the presence of in-situ foam in porous media (Figure 17 and 18). The formation of in-situ foam bank between displacing air and displaced liquid reduced significantly the channeling effect and the gravity override of air, as well as the permeability of porous media to air. Sharma et al. [65] have shown that the foam in sandpack improved the fluid displacement efficiency to a much greater extent as compared to Berea cores. In sandpacks, the fluid displacement efficiency increased at low ΔP until a maximum occurred at a certain critical pressure, then it continuously decreased as the pressure increased further. However, a linear decrease in fluid displacement efficiency was observed for Berea cores.

In order to study the effect of chain length compatibility of surfactant solutions on fluid displacement in porous media, pure surfactant (i.e. SDS) and commercial surfactants (i.e. Stepanflo 40 and Suntech IV) with various alkyl alcohols were used as mixed foaming agents. Stepanflo 40 was chosen because the highest fluid displacement efficiency was found by using it in a series of commercial surfactants, Stepanflo 20, 30, 40, 50, 60, 70, and 80. Figure 19 (a) and 19 (b) shows the results of the breakthrough time and fluid displacement efficiency as a function of chain length of alkyl alcohols. It was observed that the breakthrough time and fluid displacement efficiency were significantly influenced by the chain length of alcohols added in Stepanflo 40 surfactant solution. But there was little effect on Suntech IV by adding alkyl alcohols.

For commercial surfactant Stepanflo 40 (0.05% wt./vol.), the fluid displacement efficiency increased from 84.5% to 90% with increase of hydrocarbon chain length of alcohols from C_8 to C_{10}, respectively. There was almost the same effect on fluid displacement by adding 1-decyl or 1-dodecyl alcohol in Stepanflo 40 solution. Interestingly, the efficiency became lower (78.3%) by adding $C_{16}OH$ as compared to the system without adding any alcohol (84.2%). The breakthrough time was also found to be maximum when 1-decyl or 1-dodecyl alcohol was added in the Stepanflo 40 solution. Although the major components and molecular structure of Stepanflo 40 are unknown, it appears that the chain length of alkyl alcohols does affect the molecular packing at the air/liquid interface and surface properties of foaming solutions as well as fluid displacement efficiency in porous media.

For Suntech IV, the major components are n-C_{15-17} + Toluene and its average molecular weight is 325. Since the number of carbon atoms in Suntech IV molecule is from 22 to 24 and the structure of molecules is not a straight hydrocarbon chain, the addition of long chain alcohols exhibits little effect on the surface properties of mixed surfactants and the fluid displacement efficiency in porous media. Experimental results indeed showed that the fluid recovered by Suntech IV with various alkyl alcohols was the same or even lower than that of Suntech IV without adding any alcohol. Figure 20 [63] summarizes various interfacial properties of foaming agents and microscopic characteristics of foams in relation to porous media

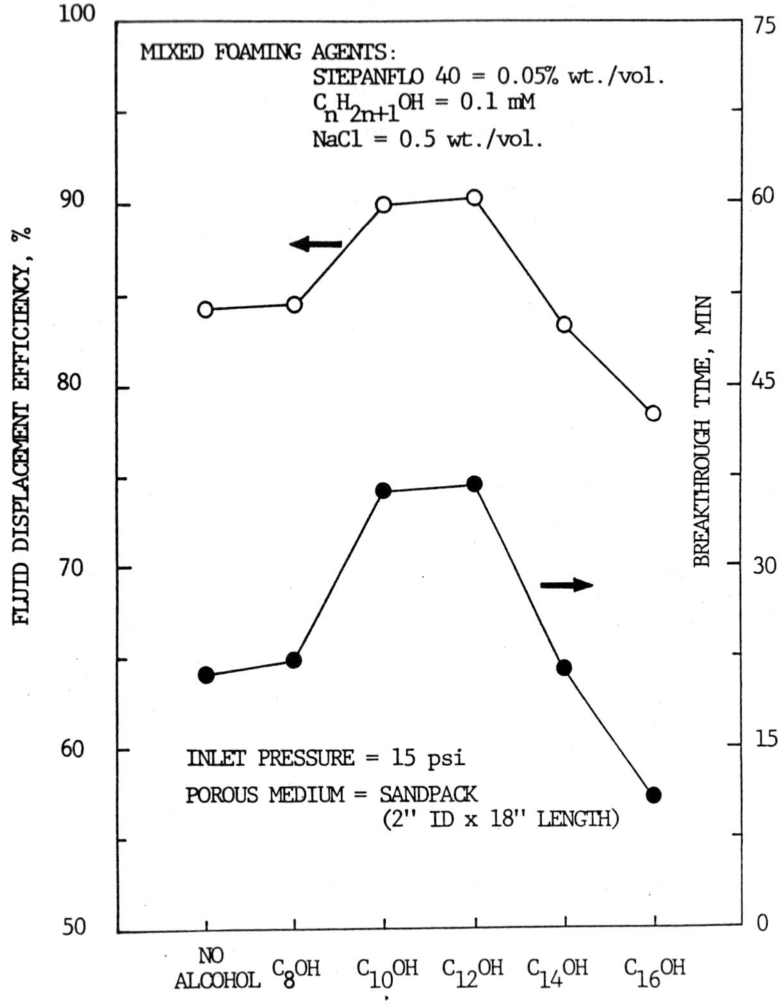

EFFECT OF STEPANFLO 40 WITH DIFFERENT ALKYL ALCOHOLS ON FLUID DISPLACEMENT EFFICIENCY AND BREAKTHROUGH TIME

Figure 19(a).

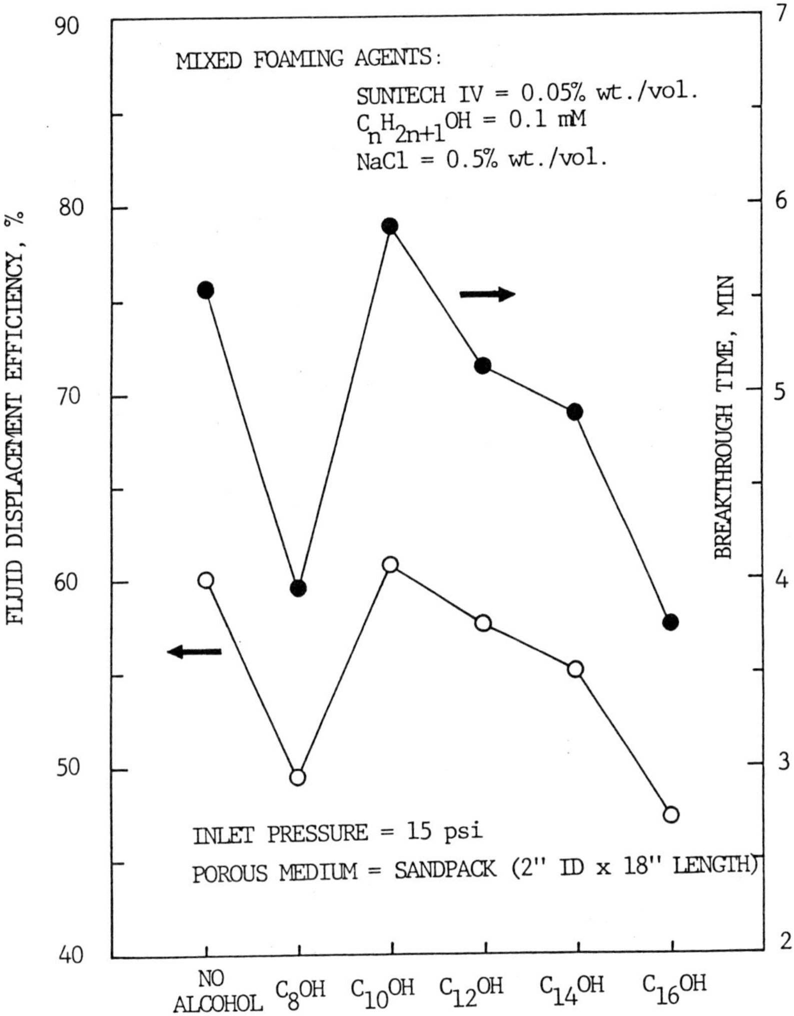

EFFECT OF SUNTECH IV WITH DIFFERENT ALKYL ALCOHOLS
ON FLUID DISPLACEMENT EFFICIENCY AND BREAKTHROUGH TIME

Figure 19(b).

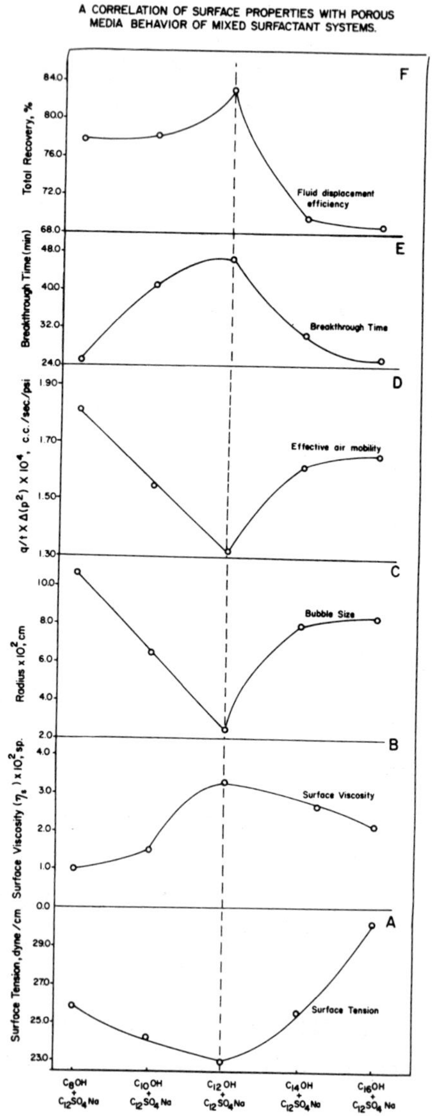

Figure 20. A correlation of surface properties with porous media behavior of mixed surfactant systems.

behavior. These results established a strong correlation of the fluid displacement efficiency with microscopic, surface, and molecular characteristics. All measured properties show characteristic maxima or minima at the equal chain length of two surfactants of the foaming system.

c. Fluid Displacement in Heterogeneous Porous Media with Foaming Agents

In order to determine the effect of in-situ foaming on blocking high permeability porous media, the sandpacks containing a non-porous (e.g. stainless steel) core or a porous (e.g. Berea core) core were prepared. The results of experiments without foam indicate that about 20% fluid can be recovered from the low permeability porous medium (i.e. Berea core). This implies that the most of air passes through high permeability porous medium and only a small portion of air passes through low permeability zone. The fluid displacement efficiency as a function of surfactant concentration is shown in Figure 21. It is clear that the concentrations of Suntech IV up to 0.075% (wt./vol.) have little effect on fluid displaced from low permeability zone (i.e. Berea core) in the heterogeneous porous media. As the concentration of surfactant was increased, a sharp increase in fluid displacement efficiency from low permeability zone was observed up to 0.15% concentration and beyond this concentration the fluid displacement did not increase significantly.

Assume that foams lose their macroscopic properties in channels which are smaller than ten bubble diameters. Based on this assumption, Dietz [78] reported that for the foams with quality higher than 0.8, the foam-drive is meaningful only if the permeabilities of porous media exceed 800 millidarcy. The permeability of Berea core used in this experiment is about 275 millidarcy. Experimental

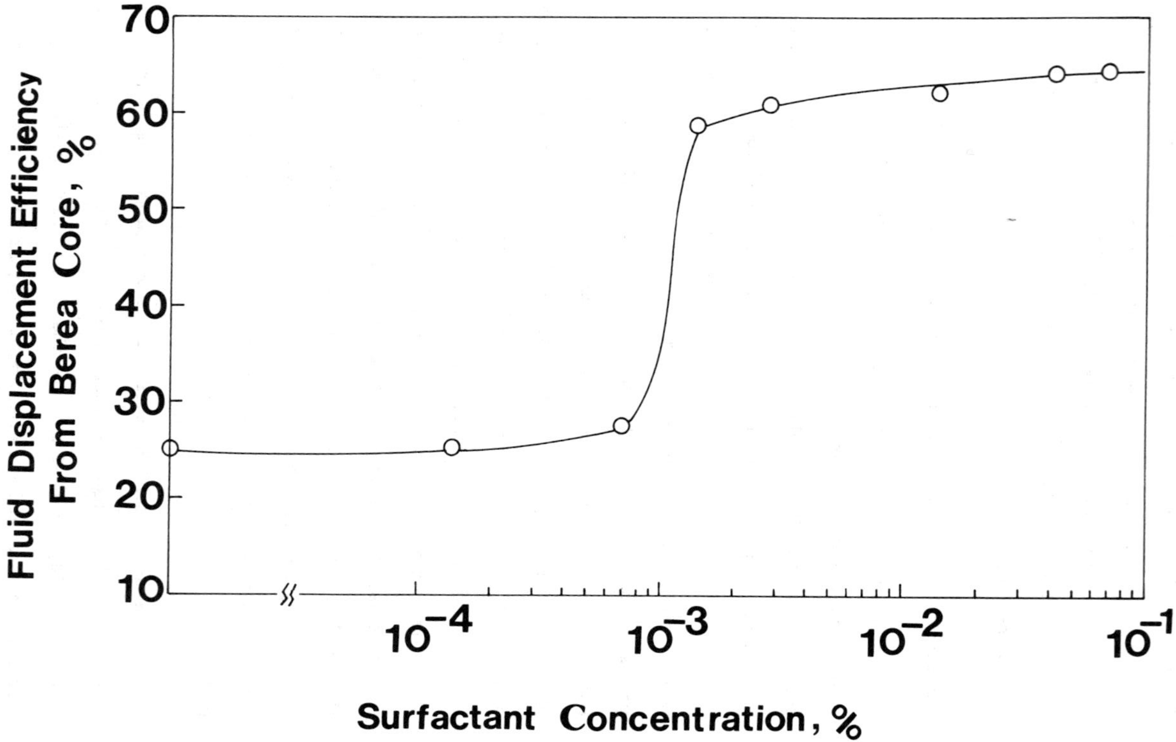

Figure 21. Fluid displacement efficiency from low permeability zones of a heterogeneous porous medium versus concentration of Suntech IV.

results showed that foams were generated in-situ and "flowed" through a portion of low permeability zone. This implies that a two-phase flow of "foam" exists in porous media and this flow can not be described by quasi-single-phase flow of a foam showing the same flow behavior as measured outside the porous media.

It appears from these results that the foam generated in-situ by Suntech IV concentration up to 0.075%, is not able to provide enough resistance to air flow. Therefore, the air passes through the easy-go channels of the porous media up to the concentration of 0.075%. Beyond this concentration, the surfactant films are stable enough to provide resistance to the air flow as well as produce large number of lamellae to displace fluid from additional pores. The suntech IV concentration close to CMC can enhance fluid displacement efficiency from 25% to 60% from low permeability porous media which cannot be recovered without the foam. This improvement in fluid displacement efficiency in the presence of foam suggests that the foams can be used to divert gas or steam from high permeability zones in heterogeneous porous media.

d. Heavy Oil Recovery by Foam Flooding

Residual oil remains in a reservoir after primary and secondary recovery because of capillary, viscous, and gravity forces. The capillary number, which correlates capillary and viscous forces, can be increased by 3 to 4 orders of magnitude by lowering the interfacial tension or reducing the viscosity of extremely viscous oils. Gravitational forces are expressed in terms of densities. Volumetric sweep efficiency will be decreased due to the density difference between oil, water, and injected fluids. Residual oil may be trapped either in the lower part of the reservoir or not contacted by an injected water-base fluid. The ratio of the mobility of the driving fluid to the fluids being displaced plays an important role in sweep efficiency. The sweep efficiency is significantly decreased as this ratio is much greater than 1. Many studies have shown that the mobility of injected gas can be effectively lowered by injecting foam. A favourable gas-oil mobility ratio is around 1.

The heavy oil recovery in porous media was measured by air flooding, steam foam flooding [79], and direct foam injection method [80] (foam generated outside porous media then injected) after the surfactant solution flooding. Experimental results showed that only 2% and 12% additional oil recovered by air foam flooding and steam foam flooding, respectively. The steam injection presumably decreased the viscosity of heavy oil and generated more foam in porous media than air injection, which favors tertiary oil recovery. At room temperature, if foams were directly injected into the porous media after surfactant flooding, then more than 20% additional oil could be obtained [80].

The economics of oil recovery methods are mainly dependent on the amount of oil that can be recovered and the cost of the injected fluids. The method of surfactant-polymer flooding has been successfully used for oil recovery by reducing the interfacial tension of oil ganglia. However, large amount of polymer solution (usually 50%-100% pore volume) has to be injected for the mobility control. In order to maximize the economics of surfactant-polymer flooding method, a novel concept is suggested so that a small slug of the expensive polymer is required. The method, namely, surfactant-polymer-foam flooding (Figure 22), has been proposed and the preliminary results showed that about 55% heavy oil recovery was attained

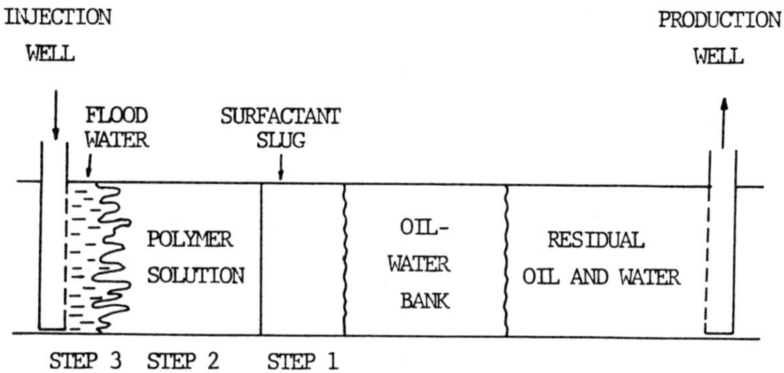

Step 1: inject surfactant-fluid bank (approx. 5-20% pore volume)
Step 2: inject polymer-brine bank (approx. 50-100% pore volume)
Step 3: inject chase-water bank

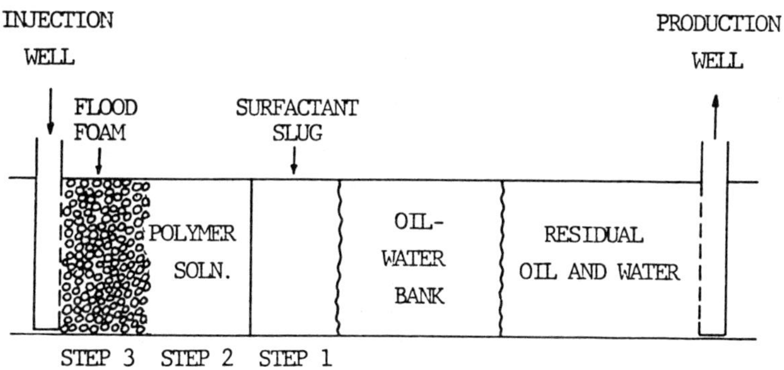

Step 1: inject surfactant-fluid bank (approx. 10% pore volume)
Step 2: inject polymer-brine bank (approx. 10% pore volume)
Step 3: inject foam bank

Figure 22. Schematic diagram of the surfactant-polymer-foam flooding in comparison with the surfactant-polymer flooding.

[80]. If foam stability is further improved and the inlet pressure of injected foam is carefully controlled, the economically acceptable rates of heavy oil recovery should result.

ACKNOWLEDGEMENTS

The authors wish to express their sincere thanks and appreciation to the ALCO Fundation and the consortium of the following Industrial Associates for their generous support fo the University of Florida Enhanced Oil Recovery Research Program: 1) Alberta Research Council, Canada, 2) American Cyanamid Co., 3) Amoco Production Co., 4) Atlantic-Richfield Co., 5) BASF-Wyandotte Co., 6) British Petroleum Co., England, 7) Calgon Corp., 8) Cities Service Oil Co., 9) Continental Oil Co., 10) Ethyl Corp., 11) Exxon Production Research Co., 12) Getty Oil Co., 13) Gulf Research and Development Co., 14) Hoechst Corporation, West Germany, 15)Marathon Oil Co., 16) Mobil Research and Development Co., 17) Nalco Chemical Co., 18) Phillips Petroleum Co., 19) Shell Development Co., 20) Standard Oil of Ohio Co., 21) Stepan Chemical Co., 22) Sun Oil Chemical Co., 23) Texaco, Inc., 24) Union Carbide Corp., 25) Union OIl Co., 26) Westvaco, Inc., 27) Witco Chemical Co., 28) the Department of Energy, and the University of Florida.

Thanks are also expressed to the former students and post-doctoral associates of the UF oil recovery research program, whose contributions form the major foundation of this review paper.

REFERENCES

1. J.J. Tabor, "Research on Enhanced Oil Recovery: Past, Present, and Future" in Surface Phenomena in Enhanced Oil Recovery, p.13-53, ed. by D.O. Shah, Plenum Publication Co., N.Y. (1981).

2. W.R. Foster, J. Pet. Tech. (Feb. 1973), 205.

3. V.K. Bansal, and D.O. Shah, in Micellization, Solubilization, and MIcroemulsions, V.I, p.87-113, ed. by K.L. Mittal, Plenum Press, N.Y. (1977).

4. R.A. Fulcher, Jr., J. Pet. Tech. (Feb. 1985), 249.

5. D.T. Wasan, paper SPE 8327 presented at the 54th Annual Fall Tech. Conference and Exhibition of SPE of AIME at Las Vegas, Nevada, Sept 23-26, 1979.

6. D.T. Wasan, S.M. Shah, N. Aderangi, M.S. Chan, and J.J. McNamara, Soc. Pet. Eng. J. (Dec. 1978), 409.

7. D.O. Shah, "Fundamental Aspects of Surfactant-Polymer Flooding Process", The paper presented at the European Symposium on Enhanced Oil Recovery at Bournemouth, England, Sept. 21-23, 1981.

8. W.S. Gogarty, Soc. Pet. Eng. J. (June 1967), 161.

9. H.L. Chang, and H.M. Al-Rikabi, J. Pet. Tech. (July, 1978), 1055.

10. H.J. Dietzel, Soc. Pet. J. (Feb. 1985), 9.

11. M.Y. Chiang, and D.O. Shah, paper SPE 8988 presented at the SPE 5th Intl. Sympo. on Oilfield and Geothermal Chemistry, Stanford, CA, May 28-30, 1980.

12. K.S. Chan, Ph.D. Dissertation, University of Florida (1978).

13. S.P. Trushenski, in Improved Oil Recovery by Surfactant and Polymer Flooding, ed. by D.O. Shah and R.S. Schechter, Academic Press, Inc., N.Y. (1977)

14. W.B. Gogarty, J. Pet. Tech. (Aug. 1978), 1089.

15. W.B. Gogarty and W.C. Tosch, J. Pet. Tech. (Dec. 1968), 1407.

16. L.W. Holm, J. Pet. Tech. (Dec. 1971), 1475.

17. L.E. Scriven, Nature (1976), 263, 123.

18. L.E. Scriven, in Micellization, Solubilization and MIcroemulsion, ed. by K.L. Mittal, V.II, p.877, Plenum Press, N.Y. (1977).

19. C. Ramachandran, S. Vijayan and D.O. Shah, J. Phys. Chem. (1980), 84, 1561.

20. R. Hwan, C.A. Miller, and T. Fort, Jr., J. Colloid Interface Sci. (1979), 68, 221.

21. K.J. Shinoda, J. Colloid Interface Sci. (1967), 24, 4.

22. K.J. Shinoda, and H. Saito, J. Colloid Interface Sci. (1968), 26, 70.

23. S. Friberg, J. Lapczynska and G. Gilberg, J. Colloid Interface Sci. (1976), 56, 19.

24. C.A. Miller, R. Hwan, W.J. Benton and T. Fort, Jr., J. Colloid Interface Sci. (1977), 61, 554.

25. G.R. Glinsman, Paper SPE 8326 presented at the 54th Annual Fall Technical Conference and Exhibition of the SPE of AIME in Las Vegas, Nevada, Sept. 23-26, 1979.

26. M.Y. Chiang, Ph.D. Dissertation, University of Florida (1978).

27. W.C. Hsieh and D.O. Shah, Paper SPE 6594 presented at the 1974 SPE AIME Intl. Symposium on Oilfield and Geothermal Chemistry, LaJolla, CA, June 27-28, 1977.

28. R.L. Reed and R.N. Healy, Soc. Pet. Eng. J. (June 1984), 342.

29. S. Vijayan, C. Ramachandran, H. Doshi, and D.O. Shah, in "Surface Phenomena in Enhanced Oil Recovery", D.O. Shah ed., pp.327-376, Plenum Publishing Co., N.Y. 1981.

30. F.I. Iskanderani, Ph.D. Dissertation, University of Florida (1985).

31. A.C. Lam, Soc. Pet. Eng. J. (Oct. 1983), 781.

32. J.A. Davis, Jr., J. Pet. Tech. (Dec. 1968), 1415.

33. H.K. Lee and D.O. Shah, Paper presented at ACS Enhanced Oil Recovery Symposium, Atlanta, GA, June 15-18, 1986.

34. J.E. Desnoyers, M. Sarbar an A. Lemieus, Canadian J. of Chem. Eng. (Oct. 1983), 81, 680.

35. H.K. Lee and D.O. Shah, paper to be presented at 6th Intl. Symposium, New Delhi, India, Aug. 18-22, 1986.

36. S.S. Marsden and S.K. Khan, Soc. Pet. Eng. J. (March 1966), 17.

37. R.Sydney, Ind. Eng. Chem. (1969), 61, 10, 48.

38. R. Lemlech, Ind. Eng. Chem. Fundam. (1978), 17, 2, 89.

39. D.L. Holcomb, Ed Callaway, and L.L. Curry, Soc. Pet. Eng. J. (August 1981), 410.

40. S.H. Raza and S.S. Marsden, Soc. Pet. Eng. J. (Dec. 1965), 295.

41. G.G. Bernard, L.W. Holm, and W.L. Jacobs, Soc. Pet. Eng. J. (Dec. 1965), 295.

42. J.T. Patton, S.T. Holbrook, and W. Hsu, Soc. Pet. Eng. J. (June 1983), 456.

43. W.R. McEntee and K.J. Mysels, J. Phy. Chem. (1969), 73, 9, 3018.

44. S. Frankel and K.J. Mysels, J. Phy. Chem. (1969), 73, 9, 3028.

45. T.O. Sibree, Trans. Faraday Soc. (1943), 30, 325.

46. S.R. Grundmann and D.L. Ford, J. Pet. Tech. (March 1983), 597.

47. R.E. Blauer and C.A. Kohlhaas, paper SPE 5003 presented at the 1974 SPE Annual Meeting, Houston, Oct. 6-9, 1974.

48. D.L. Holcomb, Drilling (Jan. 1982), 47.

49. R. Montman, World Oil (June 1982), 171.

50. W.M. Harms and J.S. Febus, J. Pet. Tech. (June 1985), 1049.

51. W.E. Warnock, Jr., P.C. Harris, and D.S. King, J. Pet. Tech. (Jan. 1985), 80.

52. M.K. Sharma, D.O. Shah, and W.E. Brigham, paper SPE 10612 presented at the 1982 SPE Sixth International Symposium on Oilfield and Geothermal Chemistry, Dallas, Texas, Jan. 25-27, 1982.

53. F.G. Helfferich, Soc. Pet. Eng. J. (Feb. 1981), 51.

54. C.C. Mattax and J.R. Kyte, Oil and Gas J. (Oct. 1961), 115.

55. J.G. Roof, Soc. Pet. Eng. J. (1970), 85.

56. J.L. Eakin and W.E. Eckard, Pet. Eng. (July 1966), 71.

57. R.F. Mast, paper SPE 3997 presented at SPE-AIME 47th Annual Fall Meeting, San Antonio, Texas, Oct. 8-11, 1972.

58. L. Minssieux, J. Pet. Tech. (Jan. 1974), 100.

59. J.C. Slattery, AIChE J. (1979), 25, 2, 283.

60. J.L. Shelton and L. Yarborough, J. Pet. Tech. (Sept. 1977), 1171.

61. R.E. Dilgren, A.R. Deemer, and K.B. Owens, paper SPE 10774 presented at the Regional Meeting of the SPE held in San Francisco, CA, March 24-26, 1982.

62. S.H. Raza, Soc. Pet. Eng. J. (Dec. 1970), 328.

63. M.K. Sharma, D.O. Shah, and W.E. Brigham, Ind. Eng. Chem. Fundam. (1984), 23, 2, 213.

64. M.K. Sharma and D.O. Shah, paper IECEC 839088 presented at the 18th Intersociety Energy Conversion Engineering Conference, August 21-26, 1983.

65. M.K. Sharma, D.O. Shah, and W.E. Brigham, AIChE J. (1985), 31, 2, 222.

66. D.O. Shah and S.Y. Shiao: "The Chain Length Compatibility and Molecular Area in Mixed Monolayers", Advances in Chemistry Series (1975), 144, 153.

67. R. Lemlich, J. Colloid Interface Sci. (1978), 64, 107.

68. T.F. Ling and D.O. Shah, paper to be presented at 6th International Symposium on Surfactants in Solution, New Delhi, India, August 18-22, 1986.

69. D.O. Shah, J. Colloid Interface Sci. (1970), 37, 744.

70. D.O. Shah and C.A. Dysleski, J. Am. Oil Chem.Soc. (1969), 46, 645.

71. A. Monsalve and R.S. Schecter, J. Colloid Interface Sci. (1984), 97, 327.

72. M.K. Sharma, D.O. Shah, and W.E. Brigham, SPE Reservoir Engineering (May 1986), 253.

73. D.L. Holcomb, paper SPE 9033 presented at the SPE Rocky Mountain Regional Meeting, Casper, Wyoming, May 14-16, 1980.

74. S.T. Holbrook, T.P. John, and W. Hsu, paper SPE/DOE 9809 presented at the 1981 SPE/DOE Second Joint Symposium on Enhanced Oil Recovery of the Soc. of Pet. Eng., Tulsa, Oklahoma, April 5-8, 1981.

75. L.W. Holm, Soc. Pet. Eng. J. (Dec. 1968), 359.

76. R.A. Albrecht and S.S. Marsden, Soc. Pet. Eng. J. (March 1970), 51.

77. M.K. Sharma, T.F. Ling, and D.O. Shah, paper presented at the 1984 SPE Forum on Foams, Jackson Hole, Wyoming, August 12-17, 1984.

78. D.N. Dietz, J. Bruining, and H.B. Heijna, J. Pet. Tech. (May 1985), 921.

79. M.K. Sharma and D.O. Shah, J. Am. Oil Chem. Soc. (1984), 61, 3, 585.

80. T.F. Ling, H.K. Lee, and D.O. Shah, paper to be presented at ACS Meeting in Houston, Texas, Nov. 19-21, 1986.

Industrial Applications of Nonionic Surfactants

By M. Hellsten
BEROL KEMI AB, S-44401 STENUNGSUND, SWEDEN

INTRODUCTION

The class of nonionic surfactants has almost become synonymous with the group of ethylene oxide based surfactants without any ionic group.

There certainly are a number of other routes to produce nonionic surfactants, of which those based on sugars and sugar alcohols and glycerol are the most important and are widely used in the food and animal feed industries. Still, for most other industrial applications practically only the ethylene oxide based nonionic surfactants will be considered due to their favourable cost-performance ratio. This type of surfactants were developed during the twenties at the IG FARBEN-INDUSTRIE in Germany and started to spread around the world immediately after the end of the second world war.

To get a perspective on the role of nonionics among all surfactants I will refer to Table 1 giving the production of non-soap surfactants in the USA, W Europe and Japan for 1982. It can be seen that the nonionics represent 30 % and that among them the fatty alcohol ethoxylates is the biggest group with more than 40 %.

This in turn is most probably due to the increasing concern about the slow biodegradation of the next largest group, the alkyl phenol ethoxylates.

Table 1 (1)

PRODUCTION OF NON-SOAP SURFACTANTS 1982
in the USA, W Europe and Japan

	1000 tons	%
FATTY ALCOHOL ETHOXYLATES	500	13
ALKYL PHENOL ETHOXYLATES	310	8
FATTY ACID ESTERS	280	7
FATTY ACID ALKANOLAMIDES	100	3
ALL NONIONICS	1190	30
CATIONICS	310	8
ANIONICS	2420	62
ALL NON-SOAP SURFACTANTS	3920	100

OVERALL VIEW ON THE INDUSTRIAL USE NONIONIC SURFACTANTS

It is popular today to talk about a helicopter view on certain areas and I would therefore like to start with some reasonably true figures for the industrial consumption of surfactants of all kinds during 1982, Table 2. The textile and fibre area is very dominating, taking about 30 % of all surfactants.

I have also made an attempt to separate out the industrial consumption of nonionics and the result is given in Table 3.

This table is naturally much more of a guesswork due to the lack of reliable information but the orders of magnitude should, however, be right.

In this table the textile and fibre field is still more dominating with about half of the total consumption, followed by the food industry taking care of 20 %. Paints and polymers resp. plant protection follow with 9 resp. 6 % of consumption.

Table 2 (1)

SURFACTANT CONSUMPTION FOR INDUSTRIAL USES 1982 in the USA, W Europe and Japan.

	1000 tons	%
TEXTILES AND FIBRES	750	29
COSMETICS AND PHARMACEUTICALS	300	12
MINING, FLOTATION, OIL RECOVERY	300	12
FOOD	200	8
PAINTS, POLYMERS	200	8
METAL WORKING	130	5
CELLULOSE, PAPER	100	4
PLANT PROTECTION	100	4
BUILDING	50	2
LEATHER, FURS	50	2
VARIOUS INDUSTRIES	400	16
TOTAL	2580	100

Table 3

ESTIMATED CONSUMPTION OF NONIONICS FOR INDUSTRIAL USES 1982 in the USA, W Europe and Japan

	1000 tons	%
TEXTILES AND FIBRES	380	48
COSMETICS AND PHARMACEUTICALS	30	4
MINING, FLOTATION, OIL RECOVERY	30	4
FOOD	150	19
PAINTS, POLYMERS	70	9
METAL WORKING	10	1
CELLULOSE AND PAPER	20	3
PLANT PROTECTION	45	6
BUILDING	5	1
LEATHER, FURS	10	1
VARIOUS INDUSTRIES	40	5
TOTAL	790	100

APPLICATION AREAS

The most comprehensive review of the industrial applications of ethylene oxide based nonionic surfactants may be found in the four monographs 2-5 written by Dr Nikolaus Schönfeldt, the first appearing 1959 and the last in 1984.

There will be no real point for me to try to cover all the various application areas for nonionic surfactants in a more detailed way, particularly as the textiles, the polymers, the mineral flotation and the coal-water fuels are covered by separate lectures during this conference.

Also the suspension concentrates in plant protection are treated by Dr Tadros as well as the enhanced oil recovery by Prof Shah.

I will thus limit myself to the following areas:

PULP AND PAPER
VISCOSE
PLANT PROTECTION
PHARMACEUTICAL PREPARATIONS
COSMETICS
FOOD AND ANIMAL FEED

NONIONIC SURFACTANTS - MOST COMMON TYPES

In order to make the following tables more understandable I have arranged the names and chemical structures of the most common types of nonionic surfactants below - together with the acronyms used in the tables 4-8.

Alkyl phenol ethoxylate
(alkyl phenol - EO)

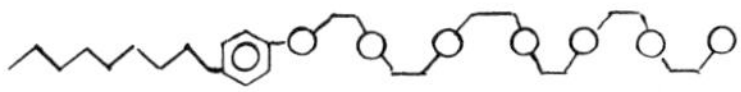

C_9H_{19} - C_6H_4 - $(EO)_n$ OH

Fatty alcohol ethoxylate
(alkyl - EO)

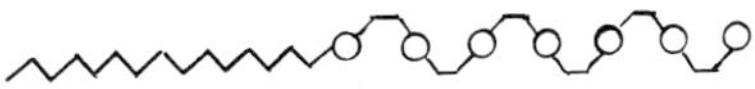

$C_{15}H_{31}$ - $(EO)_n$ OH

Fatty acid ethoxylate
(fatty acid - EO)

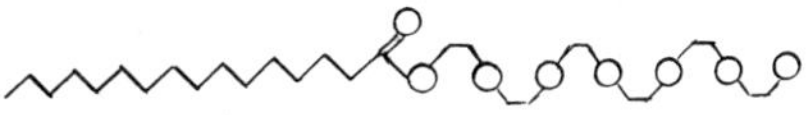

$C_{15}H_{31}CO$ $(EO)_n$ OH

Fatty amine ethoxylate
(fatty amine - EO)

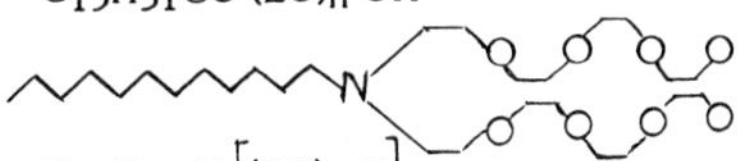

$C_{12}H_{25}N[(EO)_n H]_2$

Propylene oxide-ethyleneoxide
block polymer
(PO-EO copolymer)

H - $(EO)_n$ - $(PO)_m$ - $(EO)_n$ - H

Ethylenediamine propoxylate
ethoxylate
(polyamine - EO - PO)

H $(EO)_n$ $(PO)_p$ \ / $(PO)_p$ $(EO)_n$ H
NH_2H_4N
H $(EO)_n$ $(PO)_p$ / \ $(PO)_p$ $(EO)_n$ H

Fatty acid sorbitan ester
(fatty acid - sorbitan)

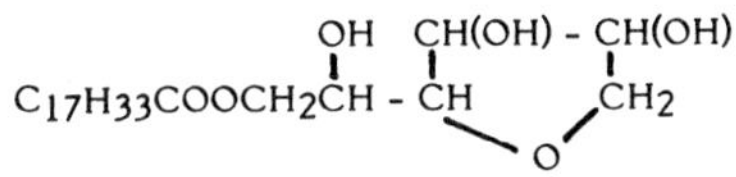

Fatty acid glyceride ethyoxylate
(fatty acid ester - EO)

$C_{17}H_{33}CO$ $(EO)_n$ - OCH_2
H $(EO)_p$ - OCH
H $(EO)_q$ - OCH_2

Ethoxylated natural oil
(natural oil - EO)

Fatty acid sucrose ester
(sugar derivatives)

PULP AND PAPER

a. Deresination

In order to get a smooth production of rayon fibres it is essential that the dissolving pulp used as raw material shall have a sufficiently low resin content, normally not more than 0.3 %.

The deresination of the pulp takes place during the hot alkali treatment at a temp. up to 120°C, a NaOH conc. of 0.5-10 % calculated on the cellulose and a pulp density of 5-15 % solids. The treatment time is normally 1-7 hours and causes a rather complete saponification of the fatty acid and resin acid esters to the corresponding soaps. During the following bleach and acid steps part of the resin and fatty acids are redeposited on the cellulose, but this can be avoided to a certain degree by the addition of a nonyl phenol ethoxylate during the alkali treatment. A normal dosage will be 1-2 kg/ton of cellulose.

The most common types of nonionic surfactants used in the pulp and paper industries are given in the table below.

Table 4

NONIONIC SURFACTANTS USED IN THE PULP AND PAPER INDUSTRY

	ALKYL PHENOL - EO	PO - EO COPOLYMER	FATTY AMINE - EO	ALKYL - EO	FATTY ACID - EO
WETTING		X			
DERESINATION	X				
DEFOAMING		X			
DEINKING	X			X	X
WAX EMULSIFYING	X		X	X	

b. Deinking

Deinking is performed according to two processes: flotation deinking which is dominating in Europe and wash deinking which is the preferred method in North America.

The first operation, the pulping stage, is the same in both processes. In this operation the waste paper is stirred in a 30-50°C solution of alkali, surfactants and hydrogen peroxide.

In the flotation process the surfactants consist of fatty acid soaps and alkoxylated fatty acids or fatty alcohols, normally in a ratio of 8:1.

In the wash process alkyl phenol ethoxylates are usually added in an amount of 2-3 kg/ton paper.

The remaining operations differ between the two processes. In the flotation process the ink that has been finely dispersed during the pulping operation is again agglomerated in soft water areas through the addition of calcium ions and the agglomerated and hydrophobic ink is removed by a flotation process.

In the wash process it is, however, essential to keep the ink in a highly dispersed and hydrophilic state because the ink will be removed by a series of washing and dewatering steps.

In the flotation process too much of the nonionic dispersants my be detrimental because they will reduce the floatability of the ink and thus reduce the whiteness of the pulp produced.

VISCOSE

The viscose industry has been in a state of recession since the late 1960s.

The tyre yarn industry was earlier a big consumer of nonionic surfactants; however, the production of tyre yarn from viscose is now steadily decreaseing, polyamide and steel taking its place.

The viscose process starts with an alkali treatment of the cellulose and in order to make the penetration of the lye more rapid the cellulose is often pretreated with a nonionic surfactant, e.g. of the block polymer type, i.e. an ethoxylated polypropylene glycol.

Various surfactants, among them also a number of nonionics, are also used in the later stages of the process as additives to the viscose itself and the spin bath thereby keeping the system clean from deposits, e.g. in the spinnerets and in the spin bath circulating system.

Fatty amine ethoxylates are also used in order to improve the physical properties of the yarn. They function by increasing the stretchability of the semiregenerated cellulose and by the increased stretch the degree of orientation in the fibre will increase.

A large amount of fatty acid polyglycol esters as well as a variety of other surfactants are used as finishing agents in order to adjust the friction between the fibres.

PLANT PROTECTION

The four main types of plant protection formulations as well as the nonionic surfactant categories used are given in the table below.[6]

Table 5

NONIONIC SURFACTANTS USED IN PLANT PROTECTION FORMULATIONS

	ALKYL PHENOL - EO	POLYAMINE - EO - PO	ALKYLAMINE - EO	FATTY ACID - SORBITAN	FATTY ACID ESTER - EO	ALKYL - EO
EMULSIFIABLE CONCENTRATES	X			X	X	X
SUSPENSION CONCENTRATES	X					X
WETTABLE POWDERS	X					X
DISPERSIBLE GRANULES	X					
ADJUVANTS	X	X	X			X

The adjuvants are intended for an enhancement of the biological activity of the formulations and probably exert their influence by improving the wetting, spreading, adhesion and penetration of the formulations under actual use conditions.

The emulsifiable and suspension concentrates will certainly be the most challenging for the formulator and as the suspensions already have been covered in another lecture I will limit myself to some viewpoints on the emulsifiable concentrates.

The user demands from these formulations that they should emulsify spontaneously, i.e. with very primitive stirring equipment, regardless of the water hardness or temperature, and that the emulsion formed should remain stable for at least two hours without any further stirring.

A typical formulation could be:

ACTIVE INGREDIENT	40-50 %
EMULSIFIER BLEND	5-10 %
SOLVENT e.g. XYLENE	ad 100 %

The emulsifier blend in most cases consists of the calcium salt of dodecyl benzene sulphonate and one or more nonionic surfactants, very often alkyl phenol alkoxylates.

As even minor changes of the solvent properties or of the concentration of the active ingredient will change the polarity of the organic phase and thus will call for an adjustment of the emulsifier blend, this means that each new formulation needs a tailor-made emulsifier system. In order to keep down the number of emulsifier blends the producers will supply a limited number of standard products which then will be mixed by the formulators to achieve the final adjustments.

The formulations are tested for spontaneous emulsification as well as emulsion stability according to standard procedures. Newer concepts like PIT (Phase Inversion Temperature) and EIP (Emulsion Inversion Point) have sofar not been very much used in the practical formulation work although the latter method seems to hold some promise as a valuable aid.

In some cases it has been possible to formulate the concentrate as a microemulsion by the use of alkyl phenol PO-EO-adducts and calcium dodecylbenzenesulphonate as the surfactant mixture. Particularly for the pyrethroid compounds this has become a favoured type of formulation. In contrast to an ordinary emulsion a microemulsion is thermodynamically stable—however only within a certain and often rather narrow temperature range.

PHARMACEUTICAL PREPARATIONS

The nonionic surfactants most commonly used for pharmaceutical preparations are shown in the table below.

Table 6

NONIONIC SURFACTANTS USED IN PHARMACEUTICAL PREPARATIONS

	EO - PO - COPOLYMERS	FATTY ACID GLYCERIDES	FATTY ACID - SORBITAN	FATTY ACID - EO	FATTY ACID ESTER - EO	ALKYL - EO	NATURAL OIL - EO
EMULSIFIERS	X	X	X	X	X	X	
SOLUBILISERS	X				X		X
GELLING AGENTS	X				X		

From this table it seems that the emulsifiers constitute the dominating area of use for the surfactants in this field. However, the most interesting area for surfactants during the last years has been the liposomes, i.e. spherical shells of surfactants which can be made to enclose a biologically active ingredient.

The difference between these liposomes and the previously known microcapsules is the much smaller size of the former, and also the better possibilities to arrange the liposome surface so that it will attach itself to a special organ in the body and there give off its active ingredient, which e.g. should be of utmost importance for the treatment of cancer with cell toxicants.

Unfortunately there seems to be a great problem in getting the liposomes stable enough for medical use.

COSMETICS

The nonionic surfactant types mostly used for the formulation of cosmetics is shown in table 7.

Table 7

NONIONIC SURFACTANTS USED IN COSMETICS

	ALKYL PHENOL - EO	FATTY ACID GLYCERIDES	FATTY ACID - SORBITAN	FATTY ACID - EO	FATTY ACID ESTER - EO	ALKYL - EO	NATURAL OIL - EO
EMULSIFIERS	X	X	X	X	X	X	X
SOLUBILISERS	X				X	X	X
HAIR CARE PRODUCTS	X				X		
DISPERSANTS	X					X	
EMOLLIENTS			X			X	
GELLING AGENTS				X	X		

The emulsifiers are dominating which seems natural with regard to the great number of O/W- and W/O-emulsions in this field.

The sorbitan esters of isostearic acid deserve a particular interest because they combine a low flow temperature with a low iodine value and thus a better resistance against rancidification.

FOOD AND ANIMAL FEED

Although rather severe restrictions are prevailing regarding additives and process aids for the food and animal feed industries there is still room for a variety of nonionic surfactants to be used which can be seen from the table below:

Table 8

NONIONIC SURFACTANTS FOR FOOD AND ANIMAL FEED

	EO - PO - COPOLYMER	FATTY ACID GLYCERIDES	FATTY ACID - SORBITAN	FATTY ACID - EO	FATTY ACID ESTER - EO	SUGAR DERIVATIVES
DEFOAMERS FOR YEAST AND SUGAR PRODUCTION	X	X	X		X	X
WETTING AGENTS FOR FRUIT PROCESSING	X					X
EMULSIFIERS FOR MILK REPLACERS		X	X	X	X	X
GENERAL PURPOSE EMULSIFIERS		X	X	X	X	X
COMPLEXING AGENTS FOR STARCH AND PROTEIN		X	X		X	
FAT CRYSTALLISATION MODIFIERS			X		X	
DISPERSANTS FOR SOLID FOODSTUFFS					X	

The defoamers and the emulsifiers are the dominating items. The emulsifiers for the milk replacers form an interesting group from a surface chemistry point of view and I would like to discuss them to some extent.

A milk replacer is a powder product, which when stirred with warm water will form a milk-like product used for the feeding of calves, pigs or sheep.

The milk replacer is produced by spraying a melted fat onto skim milk or whey powder together with soy protein or single cell protein and minor additives such as vitamins, minerals and antibiotics.

An alternative production method is to spray-dry a preconcentrated skim-milk or whey together with all the other ingredients mentioned. The mixture, which has a dry content of 55-60 %, is homogenized before spraying. Normally a monoglyceride of stearic acid is used as emulsifier.

The fat must contain a suitable blend of emulsifiers and one of the ingredients of this blend will in most cases be a lecithin product. But the lecithin alone will not be sufficient to emulsify the fat to fine droplets - it will be necessary to complement the lecithin with one of the nonionic emulsifiers shown in the table above. As the demands on this added emulsifier will depend both on the fat composition, the lecithin quality and the basic milk powder, the formulator has to test out the type and amount of added emulsifier needed to give both a good spontaneous emulsification with moderate stirring and no or very little creaming.

FINAL COMMENTS

The immediate future does not look very bright for the industries producing ethylene oxide and its derivatives. I am of course thinking on the new ethylene oxide capacity of 600,000 tons/year which has come on stream recently in Saudi Arabia and which is primarily based on a natural gas supply which locally indeed has a zero value.

The primary intention is to use this ethylene oxide for ethylene glycol production but it may later easily be converted into nonionic surfactants instead.

If this should happen the arabs will have the potential to cover almost the whole demand for nonionics in the W Europe and the USA.

This is certainly a scenario that will not come true and one good reason for this is that for many surfactant applications the better performance wins over the lower price. A deep knowledge of the customers application area will in many cases be the key factor for the business. And so I think that the timing of this conference and the initiative to organize it are very well founded and I thank you for inviting me to come here.

REFERENCES

(1) B.W.Werdelmann, Chem.Ind. 1984, 36, 321

(2) N.Schönfeldt, Oberflächenaktive Anlagerungsprodukte des Äthylenoxyds. 1959.

(3) N.Schönfeldt, Surface active ethylene oxide adducts. 1969.

(4) N.Schönfeldt, Grenzflächenaktive Äthylenoxid-Addukte. 1976.

(5) N.Schönfeldt, Grenzflächenaktive Äthylenoxid-Addukte-Ergänzungsband. 1984.

(6) G.Bognolo, European Committe of Organic Surfactants and their Intermediates, General Assembly in Rome, Italy Oct.25, 1985. Lecture "Non Detergent Applications of Surface Active Agents".

Surfactants in Textile Processing

By K. R. F. Cockett
CROSFIELD TEXTILE CHEMICAL, PO BOX 26, WARRINGTON, CHESHIRE WA5 1AB, UK

Introduction

The textile industry, whilst commonly referred to as one industry, is, in effect, a collection of a number of quite different industries. All these industries use surfactants to a greater or lesser extent. Consequently, the uses to which surfactants are put cover a very wide range. In the time available we cannot cover all these in detail; rather we will attempt an overview and look in depth at a few selected areas.

Surface active agents employed in textile processing are generally used either as processing aids to make the conversion of fibres into finished textile materials more efficient, or as means of conferring desirable effects on the finished goods. I propose to study these separately as "Processing Aids" and "Effect Chemicals".

Processing Aids

In order that we can appreciate the scope of usage of auxiliaries in processing, the following tables are given as an indication:

Table 1. Wool and wool blends.

Table 2. Cotton and cotton blends.

Table 3. Synthetics.

TABLE I

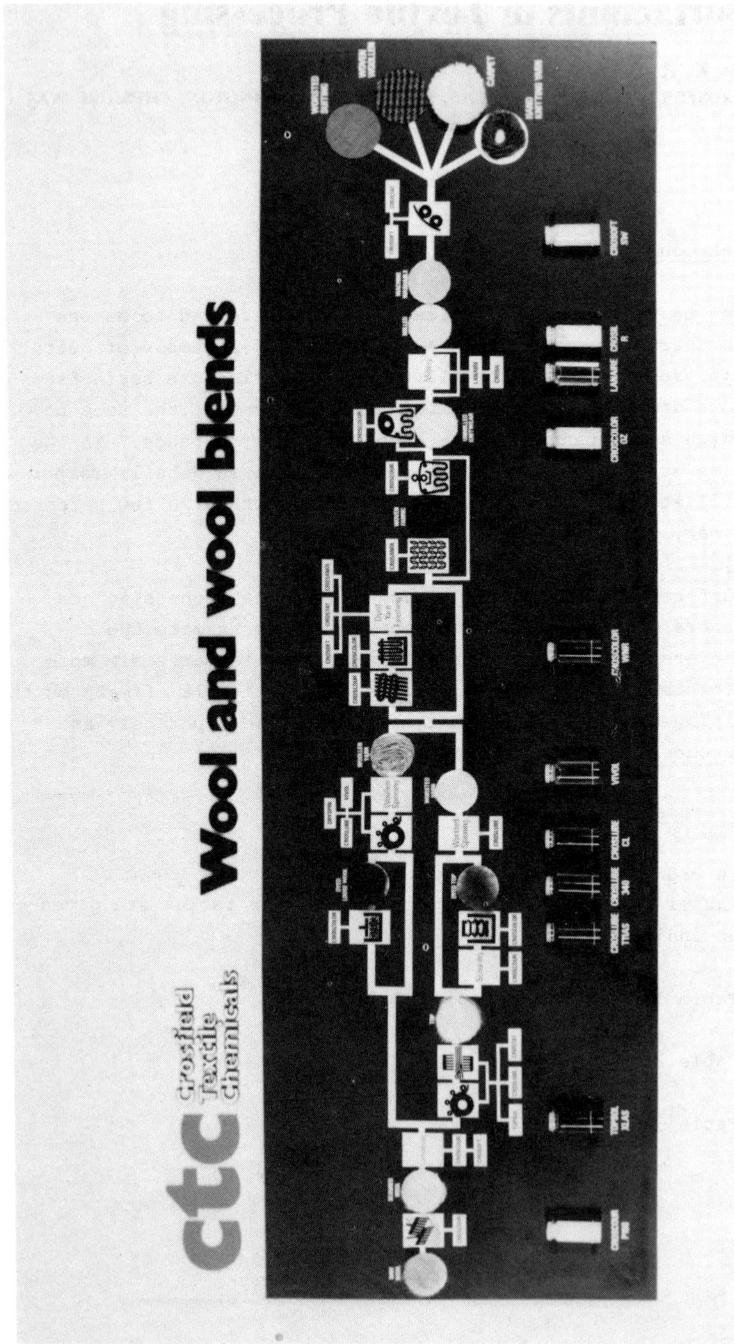

TABLE II

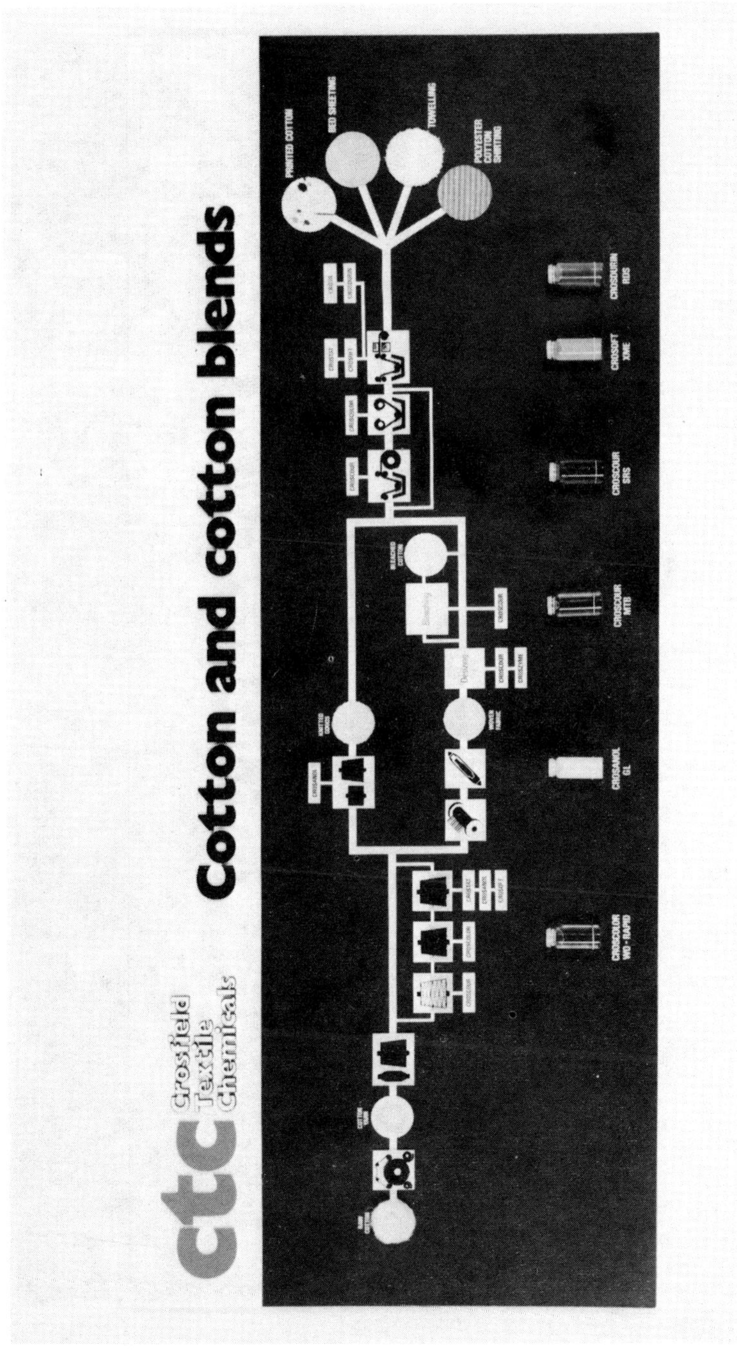
ctc Crosfield Textile Chemicals
Cotton and cotton blends
PRINTED COTTON
BED SHEETING
TOWELLING
POLYESTER COTTON SHIRTING
CROSCOLOR WO-RAPID
CROSANOL GL
CROSCOUR MTB
CROSCOUR SRS
CROSOFT XME

TABLE III

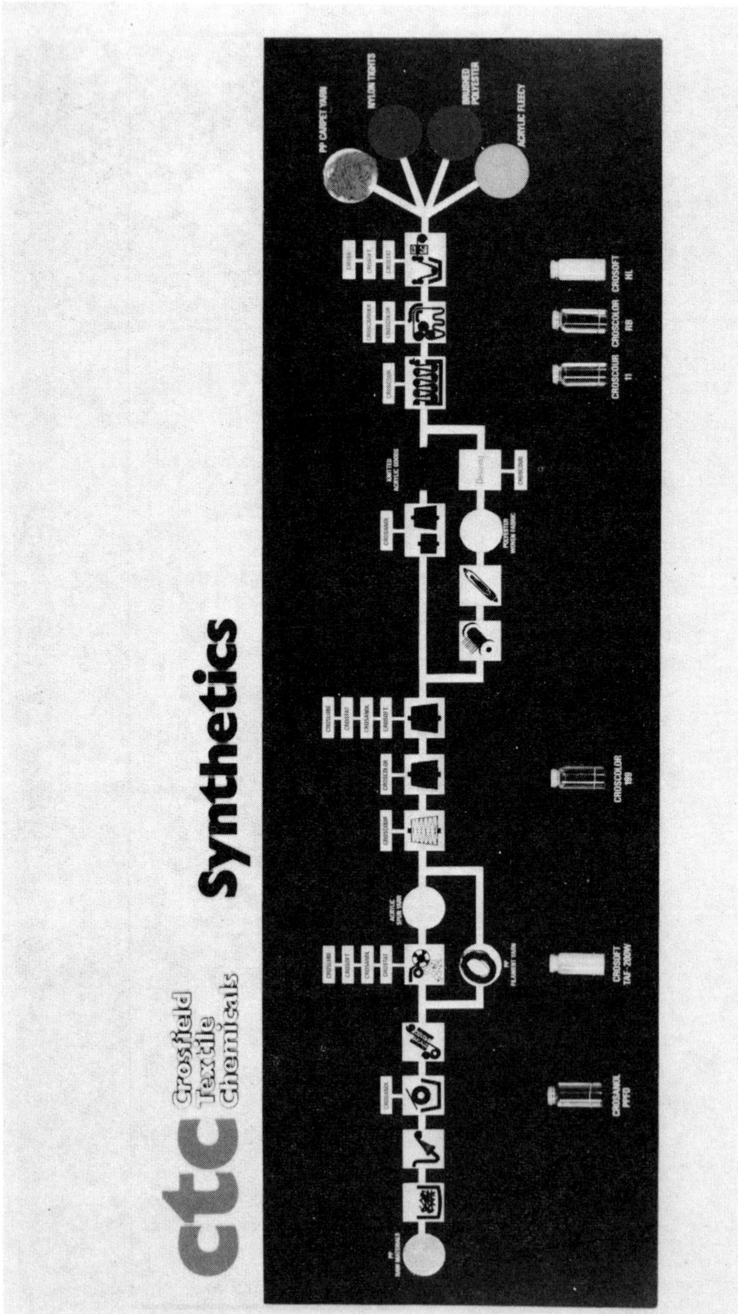
ctc
Crosfield Textile Chemicals
Synthetics
PP CARPET YARN
NYLON TIGHTS
BRUSHED POLYESTER
ACRYLIC FLEECY
KNITTED ACRYLIC GOODS
POLYESTER WOVEN FABRIC
ACRYLIC SPUN YARN
PP FILAMENT YARN
CROSANOL PPFD
CROSOFT TAF-200W
CROSCOLOR 199
CROSCOUR 11
CROSCOLOR RB
CROSOFT HL

(A) Scouring

In order to understand the scope of textile scouring it is of value to look at the variety of inherent and added contaminants that must be removed (see Table 4).

Table 4

Contaminants to be Removed

Fibre Type	Inherent	Added
Wool	Suint (complex mixture of salts)	Lubricants (see later.
	Wool grease (higher fatty acids and alcohols).	Machine oil.
	Soil	
	Degraded cellulosic material.	
	Proteinaceous contaminant layer.	
Cotton	Hemicellulose Pectins Fats and waxes Lignin Albumen.	Sizes (starch based, or synthetic), Machine oil.
	Calcium and magnesium salts.	
Synthetics	Oligomers	Lubricants
		Antistats
		Spin finishes
		Sizes.

Thus even in the field of scouring there are many different requirements.

In raw wool scouring the early usage of soap and anionic detergents has been largely replaced by the classic NP9EO and its close relations. These are frequently formulated to increase their cloud point to 50-60°C, as scouring is always undertaken above the melting point of wool grease (ca. 40°C), whilst retaining their HLB in the correct range. Generally, scouring is carried out under weakly alkaline conditions (soda ash).

Cotton scouring, on the other hand, is largely carried out in the presence of quite high (in some cases up to 100 g/l) concentrations of sodium hydroxide. This demands quite complicated formulations that give rapid wetting, good detergency, stability to caustic and, often, low foaming properties. A wide variety of chemical types is used ranging from anionic phosphate esters to nonionic alkoxylates. Products in this area are also designed to carry out functions more than simple detergency, such as peroxide stabilisers or protectors against the formation of any cellulose. An example of an actual recipe as used by the industry indicates the way in which a number of products, themselves formulations containing a variety of chemical types, are used in a treatment liquor (see Table 5).

Table 5

KIER BLEACHING

Product	Conc.	Constitution	Function
Hydrogen Peroxide (35%)	6 ml/l		Bleaching
Sodium Hydroxide	2 g/l		Degradation of seed husks and cotton wax.
Sodium Metasilicate	1 g/l		Peroxide stabiliser and alkali with buffer capability.
Organic Bleach Stabilisers	1 g/l	Organic sequestrants with nonionic and anionic wetters and detergents.	Peroxide stabiliser, wetter and penetrant.
Solvent Scour	1.5 g/l	Organic solvent with dispersing agents and anti-re-deposition chemicals.	Remove cotton wax and the degradation products of alkaline and oxidative action.
Sequestrant	0.5 g/l	e.g. Sodium heptonate	Extend the active life of peroxide to 3-4 hours at 90°C.

(B) Lubrication

Lubricants have been used for some considerable time to ease the conversion of loose fibre into spun yarn. The history of the development of lubricants for woollen spinning shows the increasing importance of synthetic surface active agents.

Table 6

The Development of Modern Lubricants for Woollen Spinning

	Product Type	Constitution	Comments
(a)	Oleines	Crude oleic acid.	Removed by addition of soda ash in saponification scour.
(b)	"Oleines"	Oleic acid/mineral oil blends.	Cheaper than (a), removed by saponification scour.
(c)	Emulsifiable Mineral Oil	Mineral oil, emulsifiers.	Cheap, readily removed in neutral scour.

(d)	Semi Neutral Oils.	Oleine, mineral oil, emulsifiers, polyglycols.	Require saponification.
(e)	High Detergent Oils	Blend of wide variety of lubricant entities, with high proportion of emulsifiers.	Water dispersible.
(f)	Polyglycols	Obvious	Only of value for coarse-count spinning. Water soluble.

Surface active agents can be used in lubricant formulations, not only to promote easy subsequent scouring, but also to give cohesion or anti-static effects. In most cases the correct surfactant pack in a lubricant formulation is crucial to ensure easy and complete subsequent removal. Inadequate scouring of e.g. carpet yarn spinning lubricant, can result in rapid soiling often shown as "tramlines" of soil on a finished carpet.

(C) Dyeing Assistants

In this area surfactants perform a great number of functions ranging from simple dispersants present in dyestuffs at point of sale, or added to the dyebath all the way through, to sophisticated amphoteric dye levellers for anionic reactive dyes. Table 7 gives an idea of the scope of such assistants.

Table 7

Dyeing Assistants

Fibre Type	Dyes Used	Functions of Dyeing Assistants
Protein e.g. wool	Acid, reactive, premetallised, chrome.	Leveller, wetting agent, protective colloids.
Cellulosic e.g. cotton	Reactive, direct, vat, azoic.	Leveller, soaping off, fixers, anti-crease.
Synthetic		
Polyester	Disperse	Dispersing agent, scour/dye carrier.
Acrylic	Basic.	Retarding, levelling.
Polyamide	Disperse, acid.	Dispersing agent, "backtan", levelling.
Blends	All	Blocking, anti-precipitant.

Levelling agents generally act by either (a) being dye substantive, thus lowering the rate of diffusion of the dye-surfactant complex, or (b) fibre substantive, thus levelling up surface irregularities or competing with dye for available sites. Examples of (a) include the ethoxylated fatty amines used as levelling agents for acid dyes. The active component of such a formulation is probably surface active and will also probably require another surfactant to hold the blend together as a coherent product. An example of (b) would be the alkylaryl ammonium halides employed as levellers in the dyeing of acrylic fibres with cationic dyes.

Dispersing agents vary from the anionic sodium naphthalene sulphonates and ligno sulphonates to nonionic alkoxylates. The former are cheap and give no levelling, the latter more expensive but do give some levelling.

Recent modifications to machinery have placed heavier demands on the surfactants, in terms of their ability not to foam even under conditions of very severe agitation. Careful blending of surfactant types can give low foaming products. Occasionally, however, silicone based anti-foams are necessary, although many users are afraid of the woven dyeings that can result when silicone anti-foam emulsions break down.

Conversely -

(D) Foamers

Recently a number of researchers have worked on using foam as a transfer medium, rather than water. The hoped-for advantage of less energy being needed for drying has been obtained although, as yet, foaming machines are not widely used. Surfactants are employed as foam producers and as foam stabilisers with systems designed to combine optimum foam density with correct half-life. The foam can be used as a means of applying dyestuffs or finishes such as crease resist resins or softeners.

Effect Chemicals

One of the most widely used effect chemical groups is that concerned with handle modification.

(a) Softeners

The improvement of handle at point of sale or during a garment's life can be achieved by a number of softener types.

Anionic softeners are rarely used even though they are highly hydrophilic and non-yellowing. Generally they are based on sulphates, sulphonates or sulphosuccinates of fatty chains such as stearyl.

Nonionic softeners are much more widely used. Typical chemical types are blends of polyethylene or wax emulsions, ethoxylated glycerides or ethoxylated castor oil.

Cationic softeners are the most popular group of softeners. Commonly they are alkyl quaternary ammonium compounds, amino amides or imidazolines.

Virtually all softener formulations are surface active. Emulsifying agents may be used to give emulsion stability, so that, for example, softeners can be padded on at the same time as crease resist resins. They may also be used to confer exhaustive properties onto compounds that have no inherent desire to exhaust.

Even though softeners are essentially effect chemicals they also give some benefits to processing. Examples that come to mind are softeners being used to improve raising and brushing, and which at the same time aid rapid sewing.

(b) Shrinkage Control

Natural fibres, like wool and cotton, shrink upon washing and need treatments to minimise that effect. The treatments used to prevent felting shrinkage of wool throw up an interesting aspect:- untreated wool has a surface free energy of around 35-40 dynes/cm and the resins that are used for covering the scales an S.F.E. of around 55. Consequently, the wool surface must be changed by chlorination to give a higher S.F.E. (70).

Recently, low free energy silicones have been used to give shrinkage control.

(c) Miscellaneous Effects

Waterproofing chemicals have long been based on fatty or paraffin waxes. Recently, though, silicones have been used to confer much greater degrees of water repellency.

A number of surfactants are applied to textiles as temporary or permanent antistats in their own right or as part of a greater formulation e.g. in lubricants or in softening agents.

Surfactants are used to formulate pesticides employed as mothproofers, via a solution in an organic solvent, into a form where they exhaust well onto wool, leaving very low residues in the exhausted application bath.

Conclusions

Almost all areas of textile processing use surfactants whether the users realise it or not. The applications we put surfactants to are rarely exclusive to the textile industry; rather they include many of those employed in a wide variety of dissimilar industries.

Formulation of Industrial Hygiene Products: Art or Science

By J. R. Sutton

DIVERSEY LIMITED, THE TECHNICAL CENTRE, MILL LANE, RIDDINGS, DERBY DE55 4BA, UK

The areas I intend to cover are:-

1. Machine dishwashing
2. Laundry detergents
3. Hard Surface cleaning

My talk will be essentially a practical overview of factors for consideration when formulating products for use in these areas. I will concentrate on surfactant selection, particularly with regard to nonionic surfactants which have indispensible properties for these applications. I do not intend to give a detailed catalogue of surfactant recommendations, but rather discuss test methods by which suitability can be assessed, since many formulation suggestions are available in various publications and manufacturers' literature.

There are a large number of surfactants available on the market and also a large amount of technical data on these materials in manufacturers' literature. Usually this information relates to standard conditions and what needs to be borne in mind is that some of the conditions of use for these applications are nothing like those used by the manufacturer for his tests. In order to simplify the selection task it is necessary to carry out tests under conditions relevant to the process of interest.

I will start by listing the advantages or useful features of nonionic surfactants that will apply to any cleaning application.

Excellent surface active properties
(wetting, emulsifying, dispersing, detergency)
Good chemical and physical stability to change of pH and temperature
Compatibility with other surfactant types
Immune to changes in water hardness
Available as 100% active liquid and solid products
Effective at low concentrations
Upper cloud points

One other factor that needs to be considered when selecting nonionic surfactants is biodegradability.

The biodegradability of nonionic surfactants has become important since the partial degradation products were found to give an unfortunate synergistic effect on the foaming properties of traces of anionic surfactants. Some nonionic surfactants, particularly the PO-EO block copolymers were found to be particularly resistant to biodegradation as were alkyl-phenol ethoxylates when the alkyl chain was branched. As a result of these factors the EEC produced a directive in 1973 that only nonionic surfactants would be permitted that gave a result of 80% biodegradability. However, they also produced a derogation which excused products in certain end uses until a certain date - on the grounds that nothing was available that was biodegradable and suitable.

The derogations of interest to us are as follows:-

a) low foaming polyalkylene oxide addition products used in industrial cleaning agents and mechanical dishwashing products.

b) alkali resistant terminally blocked and alkylaryl polyglycol ethers used in industrial cleaning agents for the food, beverage and metal working industries.

The date for this derogation to expire has been extended over the years and the latest word is that it has now been finally extended only until 1989 so it is obviously necessary for steps to be taken to start ensuring that formulations will still be acceptable after this. In any case the legislation further says.

"In the event of industrial use of the surfactants referred to in a) and b) above the user should employ the best techniques available to limit the emission of these into the aquatic environment".

In cleaning applications generally of the type we are interested in the following factors must be considered as affecting cleaning performance. These are substrate type, soil type, chemical energy (i.e. the formulation), temperature, time and mechanical energy (agitation). The contribution of the formulation can cover a wide range of effects, namely detergency (wetting, dispersion and emulsification) - to which the surfactant contributes, plus solvency, anti-redeposition, neutralisation and antistatic effects.

The low foaming nonionic surfactants as a class are unique in exhibiting very little foam above their cloud points but otherwise can retain their desired surfactant properties at these temperatures. With low foaming surfactants it should be noted that they are only low foaming above their cloud points. In fact all nonionics are lower foaming above the cloud point as can be seen below from the effect[1] of cloud point and temperature on foam generation for a normal and a low foaming nonionic in the Hamilton-Beach foam test.

Table 1

		Foam mm for 1% Solution	
	Cloud Point °C	25°C	60°C
NP + 9 E.O.	54	127	62
Commercial Low Foamer	27	107	7

Machine Dishwashing

One useful review of dishwashing is that by Mizuno[2] which explains all the intricacies of machinery and operating practices.

The need for surfactants in machine dishwashing is three-fold, namely:-

1. detergency
2. defoaming
3. rinse aid

Tackling these in order. In industrial machine dishwashing most of the detergency is provided by the alkali chemical in the wash solution - mostly sodium hydroxide or sodium carbonate with various alkaline builders plus a large contribution from the agitation of the wash jets which run at quite high pressures. These effects combine to make the use of large quantities of foaming surfactants both unnecessary and impractical since the level of foam produced would be excessive and would reduce the effectiveness of cleaning action of the jets and also fill the vigorously agitated tank with foam. The only types of surfactant that it is practical to include are very low foaming nonionics. These are often included at the level of approx. 1-3% in the products which themselves are used at the level of 1-3 grams per litre with a wash temperature of 60°C. If it is required to obtain measurements of detergency in machine dishwashing it is necessary to reproduce real conditions as closely as possible using a dishwasher and appropriate test pieces - these should be in good condition to avoid confusing results due to non-reproducible surface conditions. These test pieces should be artificially soiled with soils reproducing as closely as possible the type of soil thought to be of interest i.e. starch, proteinaceous or fatty. The use of various soils has been reported in the review on dishwashing by Mizuno[2]. The required soiling technique, whatever the nature of the soil is such that it is applied in such a thickness, and allowed to cure (air dry) to such an extent that even the best detergent does not remove it all - what is needed is a gradation of removal from the poorest to the best detergent.

With regard to measuring the remaining soil, various methods have been used varying from use of reflectometers to use of radio-tagged soils. Again I would refer you to the review by Mizuno[2] for reference to original papers for details of all the methods reported.

One standard method for measuring dishwasher detergency that has been published is by ASTM.[14] This measures the build up of soil (margarine + powdered milk + cooked wheat) after multiple repeated washes of glass tumblers. The tumblers are rated for spotting and filming on a scale of 1 to 5 (1 = no spots or film, 5 = complete coverage) after 5-15 cycles. The use of this test at 60°C has been reported to give satisfactory results for various commercial dishwasher chemicals after 15 cycles. The actual surfactants used can be just low foaming or they can have a defoaming action as well. Surfactant types which have been widely used are as follows:-

The Pluronics - which are PO-EO block copolymer

$$HO\text{-}(EO)_a\text{-}(PO)_b\text{-}(EO)_a\text{-}OH$$

Benzyl end blocked alkylphenol ethoxylates

Benzyl end blocked alcohol ethoxylates

Alcohol EO-PO block copolymers

With regard to stability of these surfactants in the alkaline environments associated with dishwashing this definitely appears to be favoured by the end capped polymers - non-capped ethoxylate surfactants are often degraded into significantly foaming products. In addition to these types of surfactants, others have been used alone and in combination with the above types including phosphate ester surfactants such as stearyl acid phosphates. It is claimed that these even suppress the foam at fairly low temperatures where the ordinary nonionic defoamers no longer work. Owing to the way that commercial dishwashers are operated there can in any case be a foaming problem in the wash tank solution. This is because the wash solution is used repeatedly (with regular additions of fresh detergent to maintain detergency) and there is a continuous build up of food soil in the solution which can consist of degraded protein and fats and have inherent high foaming properties. The need for a defoaming surfactant in the wash tank is greater in fact than for a surfactant providing detergency. The most suitable types of surfactant for this function are low foaming nonionics. What is required in the cleaning solution is one with a cloud point below that of the dishwasher wash temperature - this is usually 60°C or thereabouts, so obviously there is a large choice available.

Evaluation of surfactants for suitability with respect to foaming tendency and food soil defoaming ability can be checked by simple laboratory tests - either using the dynamic foaming test in the Ross and Miles apparatus[4] or in a simple stoppered cylinder shake test - of course carried out at the temperature of interest. The latter test consists of carefully heating a weak (50 ppm) solution of the surfactant in a dilute solution of a suitable machine dishwashing detergent base consisting of 1 g/litre sodium hydroxide, 1 g/l (tripolyphosphate) and 1 g/l hypochlorite (10% Av.Cl_2). For the defoaming version of this test 1 ml of a 10% solution of dried milk should be included. To measure the foaming/defoaming effect, 50 mls of this solution contained in a stoppered 250 ml measuring cylinder, is heated to the required temperature in a water bath and then shaken vigorously by rapidly inverting the cylinder 10 times in 5 seconds, noting the volume of foam immediately on completion of shaking and 5 seconds afterwards. In the case of evaluations from this test for food soil defoaming it is recommended that for suitable surfactants the immediate and 5 second foam levels should be as low as possible. If possible no more than 5 mls immediately and decaying as much as possible. This is quite a demanding test considering that the milk soil alone in this test solution without surfactant gives approx. 40 mls foam initially, decaying to 30mls.

I will now move onto the third function of surfactants in machine dishwashing which is as rinse aids. The features expected from a rinse aid are listed below:-

Improvement of appearance of washed ware
Reduction in drying time
Defoaming in the wash tank

Rinse Aids perform an essential task in that they help the rinse water form a continuous film on the surface of plates which then gives a good final appearance after drainage and drying. To give the continuous film, nonionic surfactants are added at a low level (50-100 ppm typically) to the rinse water, in order to lower the surface tension sufficiently to enable spreading.

The exact concentration of nonionic necessary varies with the structural type and also the point of addition to rinse water - the latter affects how well the surfactant is dispersed.

The appearance of the rinsed articles - assuming good detergency in the wash has removed all food soil and that all detergent residues have been rinsed off - depends on the presence of alien matter on the surface which can come only from the rinse water itself or from the rinse aid surfactants. With regard to the water solids, these are the hardness salts e.g. calcium and magnesium salts and it has been found using an efficient rinse aid that these can become a problem at around 250-300 ppm in the water. If too much rinse aid is used this can give a problem due to the surfactant itself depositing on the surface. Apart from this overdosing, poor dispersion of the surfactant in the rinse water can give rise to this deposition on the surfaces. Good dispersion is not easily achieved under the conditions of commercial dishwashers since the surfactant is normally injected into the hot rinse water at 80-90°C in the water line only a few inches from the rinse jets. It is thus subject to only a short mixing period under conditions of minimum solubility i.e. above its cloud point.

The criticality of appearance is most marked with glass-ware and it is recommended that comparisons of performance of surfactants is made with glasses. Ideally these should be looked at under conditions of constant illumination - indeed we have gone to the trouble of constructing a viewing black box, where the inverted glasses are illuminated from underneath with a fluorescent light, and viewed in a darkroom using black squared templates to enable spots or rather spot free areas to be counted. Obviously there is quite an art in rating appearances of washed articles in this way and what we did to introduce a little science into it, and also a little confidence in the results, is to do several replicate assessments and follow this by statistical analysis of the data. What we found was that you could not compare results between operators and that you always need to run a standard.

Drying times can be determined in a test involving wetting glass plates in 100 ppm solution of the surfactants and noting the time to dry in a draught free environment. Several replicates are required and also a statistical test to determine significance. The following results were obtained for a series of nonionics - with a confidence interval at the 99% level of 2.1 secs.

Table 2

	Drying time seconds
Pluronic L81	37.0
Pluronic L31	42.5
Coco alcohol + 5 E.O. (C_{12})	50.0

To summarise then, suitable surfactants for use in rinse aids should have the following properties.

(a) Give Good Film Forming Properties

This requires a low enough surface tension for spreading to occur. The situation is complicated with regard to correlation of lab measurements such as capillary rise since this is related to equilibrium conditions and the rinse aid situation is related to an earlier point on the surface tension - time curve. Also affecting this must be the fact that the rinse aid solution at the high temperature of interest (80°C) is in fact two-phase. A most suitable method for measuring surface tensions of surfactants under rinse aid conditions is the bubble pressure method of Sugden[5,6]. With this method which involves measurement of maximum bubble pressures for bubbles formed from two different sized tubes it is possible to conveniently measure surface tensions at high temperatures and for surface ages from 1 second upwards. The method showed, that where the surfaces remained oily after incomplete washing, a surface tension of between 30-39 dynes/cm could be needed to give a continuous film depending on the cleanliness of the surface and the type of surface contamination.

(b) Be Readily Dispersible in Hot Water

It is known that the degree of dispersion of a rinse aid surfactant in the rinse water can be very important in influencing whether a continuous film is obtained or not.

The best method we have found for assessing the dispersibility of candidate rinse aid surfactants is particle size measurement by Coulter counter. We developed a test where the surfactants of interest were first dispersed under standard conditions in a high speed blender (Atomix). In figure 1 you can see the effect of dispersion technique and droplet size distribution correlated with the appearance of the rinse aid film in a dishwasher. It can be seen that increased agitation gives smaller particle sizes and improved performance. The best performance was obtained from a cold solution that had been heated to above the cloud point.

(c) Not Cause Deposits on Plates

It is important with rinse aids that all the surfactant is well dispersed without any large droplets remaining since above a certain sizè the droplets hit and adhere to the surface being rinsed, further droplets will then coalesce with the ones already adhering and the end result after drying will be a spot or smear on the surface.

Figure 1 : for PO/EO Block Copolymer MW = 2500 EO = 20% by weight

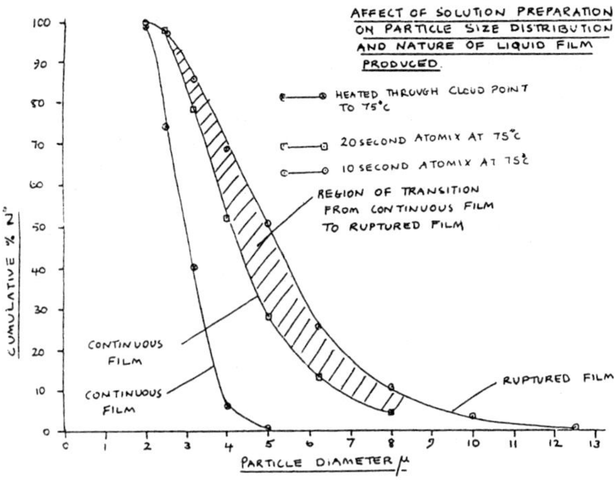

(d) Not Foam in the Machine

One of the main limitations on the formulation of an effective rinse aid is its foaming property. If it is too high foaming it will generate undesired foam in the dishwashing machine which will destroy the washing action of the machine as well as spilling onto the floor.

(e) Defoaming Action

With regard to the last desirable rinse aid feature of defoaming performance, this results from the way that dishwashers work - they usually overflow used rinse water into the wash tank so this puts the rinse aid into the wash water where it can do a useful job in defoaming the food soil, which incidentally can reduce the need for incorporating surfactants into the dishwasher detergent. In evaluating surfactants for suitability in this respect initial testing can of course be carried out by the test described earlier - These surfactants obviously need to be very low foaming in their own right and a cylinder shake test at 50°C in distilled water will aid in their selection. Ideally foam levels should be very low as mentioned for surfactants in the dishwashing detergent stage of course.

The science (or art?) of formulating rinse aids is to combine all these properties in the one surfactant or surfactant blend. The oldest rinse aid technology has used the EO-PO block copolymer surfactants but their low biodegradability has now made their use less popular. The problem of finding a suitable compound is one of conflicting properties - those compounds which are sufficiently low foaming usually are too hydrophobic to disperse sufficiently easily. Increasing the ethoxylation degree to aid dispersion causes the foam level to increase to unacceptable levels. The more biodegradable alcohol - EO-PO surfactants and other biodegradable systems which are available are less readily dispersible. This problem can be overcome by incorporation of low foaming hydrotropes such as sodium xylene sulphonate into the rinse aid product, but these can contribute to the foaming problem and they will certainly contribute to the solids load for the rinse water and may contribute to spotting.

Laundry Detergents

In the formulation of laundry detergents we can be involved with both anionic and nonionic surfactants. A lot of papers are published on the subject of formulation and efficiency of laundry detergents. The majority of items are found in the Amercian journals and the majority of these apply particularly to products for the domestic market. However, a lot of the information in these papers can be usefully applied to the institutional and industrial markets. In these applications foaming is not quite as critical as in machine dishwashing and the anionics can serve a useful function. As a rough guideline to foaming and its presence in washing systems - which are usually made alkaline with builders to sequester water hardness salts, the following general statements apply:-

1. Anionic surfactants in general produce a lot of foam, usually less in hard water but increase with increase in temperature.

2. Nonionics generally foam less than anionics although it is necessary to use low foam versions to operate without any additional foam control.

3. Addition of soap to an anionic detergent depresses foam.

4. Nonionics do not depress the foam of anionics - in fact they can give increase in foaming.

5. All types of surfactants foam more in the presence of alkaline builders.

Returning to the subject of detergency, lots of detailed studies have been carried out on fabrics but I will just give the following general summary.

The most commonly used individual constituents have the following relative detergencies[7] - all compared in the diagram below at equivalent activities (45% active) - these being average figures covering all types of soil, cloth types and water hardness levels.

Figure 2

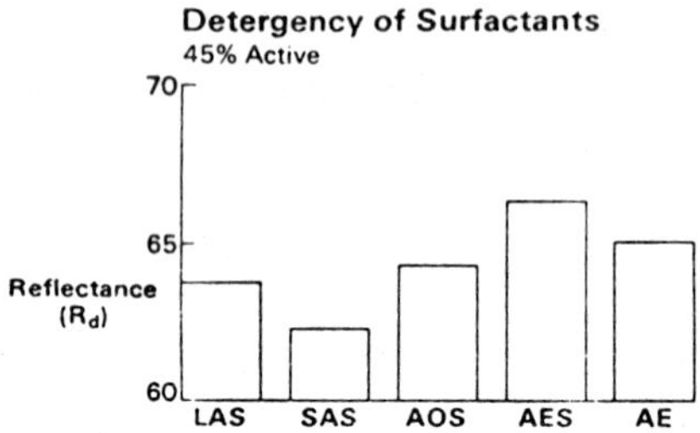

LAS = linear alkylbenzene sulphonate

SAS = secondary alkyl sulphonate

AOS = alpha-olefin sulphonate

AES = alcohol ether sulphate

AE = alcohol ethoxylate

It can be seen that the alcohol ether sulphates show the best performance but they are considerably less economical than the only slightly poorer performing nonionics. The most commonly used combination is that of alkylbenzene sulphonate plus nonionics like the alcohol ethoxylates; these have complementary properties which can result in synergistic blends as shown in the next diagram[7]. Again these values are overall averages:-

Figure 3

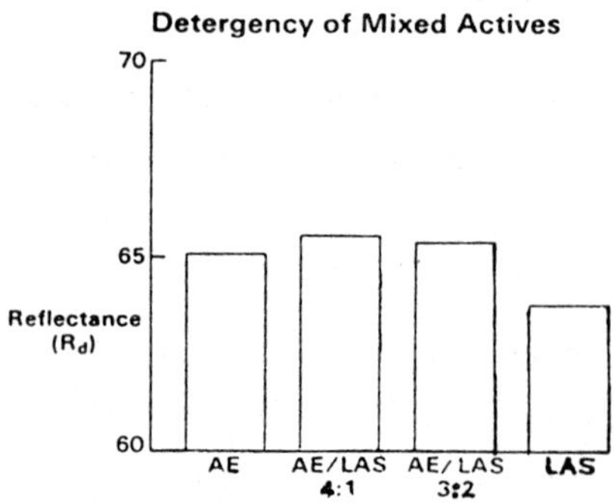

The preferred anionic component is usually the C_{12} (dodecyl) benzene sulphonate, this choice being explained by the next diagram[8] which shows the relationship between detergency and alkyl chain length (against sebum on cotton and synthetic (permapress) fibres for low and high water hardness, in the presence of phosphate builder) and alkyl chain length.

Figure 4

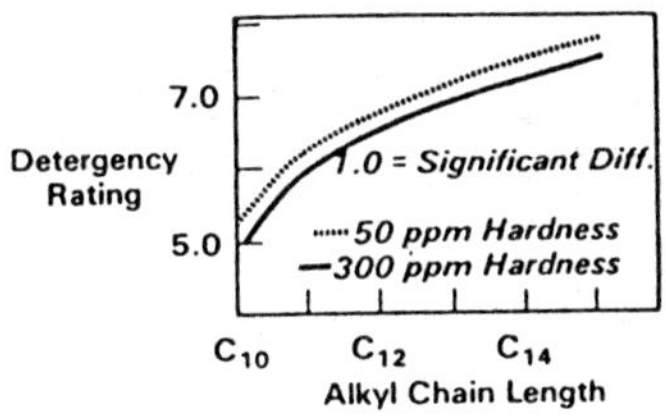

The effect of alkylate chain length on detergency performance for LAS homologs (0.20% concentration, 50 and 300 ppm hardness, sebum soil).

In considering the selection of suitable nonionic surfactants both the alcohol chain length and the extent of ethoxylation need to be considered. The effect of these two factors on detergency is shown below for a polyphosphate built system[8], the soil is sebum and the cloth the average for cotton and synthetic (dacron and permapress).

Figure 5

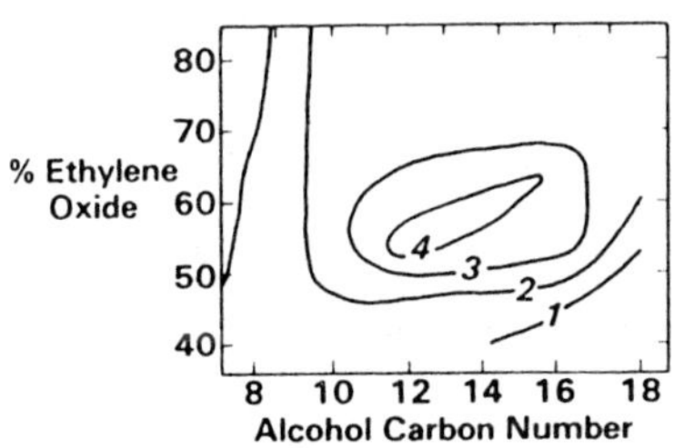

FIG. 2. Detergency optimization curves for alcohol nonionics as a function of alcohol molecular weight and percentage of ethylene oxide (10% NI, 30% STPP formulations; 0.2% concentration, 50 ppm hardness, sebum soil). (Larger isodet value = better detergency performance.)

From this it can be seen that the optimum is for a C_{12}-C_{16} alcohol with 55-65% by weight of EO. Looking at the next chart relating weight % ethylene oxide to moles ethylene oxide for different chain length alcohols it can be seen that this is approx. a 9 mole ethoxylate.

<u>Figure 6</u>

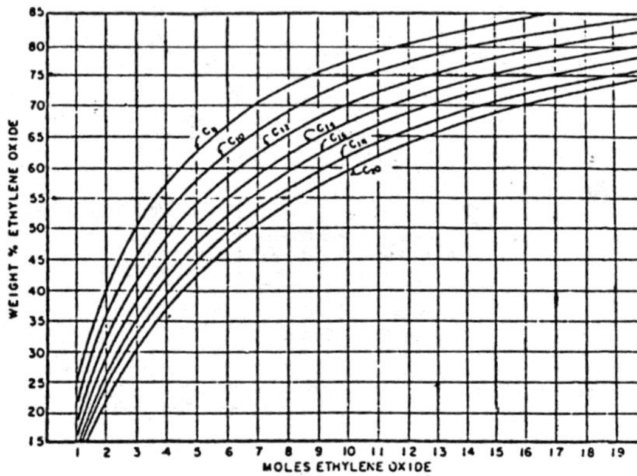

MOLES EO vs WEIGHT % EO FOR DIFFERENT ALCOHOL SUBSTRATES

With regard to relative detergencies of these surfactants against different types of soil it is generally true that alkyl-benzene sulphonates show optimum performance against particulate soil and typical body soil whereas nonionics perform better on oily soils, being particularly good at removal from synthetic fabrics. A further point to aid in surfactant selection is that the anionic surfactants are more dependent on builder action so that if the builder is weak then choice of a nonionic might be best - although of course it is probably cheaper to improve on the builder system.

The information above is probably equally relevant to powder or liquid formulations but with the formulation of liquid products (which are of increasing importance in institutional laundry products due to the facility they give for automatic dosing) where additional constraints apply. Two types of liquid formulations are used, built and non-built; the latter type would use a separate builder pack and would, of course, enable variation of the ratio of surfactant to builder to meet requirements. For the built formulations the alkylbenzene sulphonate/ nonionic surfactants combinations are used usually with the highly soluble and hydrolytically stable builder tetra potasssium pyrophosphate.

The optimum chainlength for the alkyl group in the anionic is dictated by solubility as C_{11}-C_{12}[8]. One type of product offered to the industrial markets is a separate booster pack for removing oily soil from workware i.e. overalls. This is added to a normal wash and is based on just nonionic surfactants - usually a wide range of HLB types sometimes with a small amount of water and hydrotropes to avoid gel formation and to aid solution in the wash tank. An alternative type of formulation for this same purpose uses nonionic surfactants plus a few percent of kerosene; the nonionics used serve additionally as emulsifiers for the kerosene in the product and the two materials both assist in removal of oil stains.

Laboratory investigations of detergency are usually carried out in a Terg-O-Tometer[9]. This is a battery of small scale (1000 ml) washing machines where the key variables such as water temperature, rate of agitation, detergent, water hardness and cloth/ liquid ratio can be accurately controlled. In this equipment the test pieces used are approx. 4" square pieces (swatches); you can of course prepare your own soiled swatches but it is probably easier and more reliable to purchase soiled swatches from a commercial source such as FCRA[10] or EMPA[11]

Detergency of soil removal can be measured using a colorimeter and applying the Kubelka-Munk equation[3] when % soil removal is related to the observed reflection as follows:-

$$\% \text{ soil removal} = \frac{K/S_s - K/S_b}{K/S_s - K/S_o} \times 100$$

The expression K/S is the ratio of the absorption coefficient (K) to scattering coefficient (S).

i.e. K/S_s = ratio before washing

K/S_b = ratio after washing

K/S_o = ratio of original unsoiled swatch

The values of K/S can be obtained direct from reflectance readings by using tables.

With colourless soils it is necessary to use radioactively tagged components in the soil and counting the activity before and after laundering.

Storage stability testing of products is necessary with both powder and liquid products since caking of powder products and separation of liquid products at times seems as important as detergency. Prolonged storage at low ($5^{o}C$) and high ($37^{o}C$) temperature gives information as to the long term storage characteristics. Separation of liquid components from powder products can be detected by storage on absorbent paper. With liquid products the low temperature haze point gives an indication of the minimum safe storage temperature. Additionally, liquid detergents should give good reconstitution after about three freeze-thaw cycles.

Hard Surface Cleaners

"Hard Surface" is a very wide definition of a surface, particularly in an industrial context and includes a great variety of types from trawlers to concrete floors to eggs - it excludes the areas I have covered already.

The one feature usually common in the material used in these products is alkaline salts, and strong caustic is often included, more so than in the areas I have already covered. Acidic products are also used and, as for the alkaline products these are made as concentrated as possible. In this general area the surfactant properties most studied are foam characteristics appropriate for the method of application, stability in, or on strong alkalis and acids and detergency.

The second requirement after understanding the raw materials available is understanding the intended product application i.e.

what is cleaned and how it is cleaned. The nature of the soil and the ease of corrosion of the substrate must be taken into account together with any relevant government regulations when designing a product for a specific application.

Common hard surface cleaning methods include:-

1. <u>C.I.P.</u> = clean in place

Where the cleaning solution is circulated and sprayed onto the surface to be cleaned. The dominant feature of surfactants required here is low foaming.

2. High pressure-low volume cleaning
3. Foam cleaning
4. Soak cleaning
5. Manual cleaning - bucket and mop

This last method is becoming less important with increasing labour costs. The trend is toward automated cleaning and to facilitate this the products of choice are fully built liquids - this enables pumping by a pump controlled by sensor (often conductivity) of only one cleaning product. This approach of course simplifies the cleaning routine but because the liquid products are as concentrated as the laws of solubility allow, it is necessary to understand the interaction of materials in solution. Because of this the formulation depends almost as much on stability as on performance.

For these different cleaning techniques the product compositions necessitate product characteristics not directly associated with detergency. For example where foam is detrimental, such as in CIP cleaning, defoaming surfactants are commonly used. These are usually low cloud point nonionics and it is often difficult to formulate these into stable liquid alkaline concentrates - it is necessary to use suitable hydrotropes such as sodium xylene sulphonate or one of the proprietary phosphate esters available from a number of suppliers. These, of course, can add significantly to product cost and often to the foaming characteristics as well. It is also worth bearing in mind that cleaners based on this type of system can be adequately low foaming at say 70°C but that if the temperature is allowed to drop to 50°C then foam can become a problem.

The technique of foam cleaning is often the chosen method of hard surface cleaning. Foam cleaning is most often used on vertical surfaces where a long contact time is required, often to maximise biocidal action and this can last for several minutes or more. The main disadvantage of foam cleaning is that it is effectively a low temperature process and this can restrict its efficiency. An advantage of foam cleaning is that it can readily be seen which parts of the surface have been in contact with the cleaning solution. The use of chlorinated products can produce a stability problem and it is often easier to inject the surfactant product and hypochlorite solution separately into the foaming device just before use. The highest foaming surfactants are used and these tend to be anionics but the linear alkylbenzene sulphonate is not often used due to solubility limitations. The most commonly used anionics are some phosphate esters, ether sulphates together with amphoterics and amine oxides. In these products the role of detergency is almost secondary to foam performance; various types of foam are achievable from thick, dry, luxurious to thin and wet.

Laboratory evaluation of hard surface cleaners is often difficult. There is no truly dedicated/general purpose equipment such as is available for example for fabric detergency testing or dishwashing. Various lab scale test equipment has been devised for evaluating detergencies in circulation systems with soiled panels such as microscope slides and metal coupons; use of suitably marked (e.g. dyed) soils often makes measurement of soil removal possible. However, it is by no means always true that surface cleanliness corresponds to soil removal, especially if corrosion or other alterations occur. One method recently published[12] for measuring surface cleanliness of metals by using spreading or non-spreading of solutions of known surface tension could seem to be of more general usefulness.

In the specific area of institutional hard surface cleaners i.e. for manual use on general cleaning of sinks, floors, walls etc. a wide range of surfactants are used. Popular systems include alkylphenol ethoxylates, together with a solvent such as butyl cellosolve and conventional builders. The butyl cellosolve is added to increase the grease cutting power of the cleaner. Another type leaves out the solvent and uses a mixture of nonionics that vary widely in their ethylene oxide content - usually an oil soluble low EO derivative to improve performance on oily soils and a water soluble - high EO nonionic to assist in removal of particulate soils. A recent detailed study[13] of the hard surface cleaning performance of a series of alcohol ethoxylates has shown that overall, the optimum composition for performance consists of a blend of C_6, C_8 and C_{10} alcohols ethoxylated to 50% ethylene oxide content. It was also noted that the most effective nonionic varied with the use concentration varying from C_6 for a 5% concentration to C_8-C_{10} for a 0.2% solution. The study also revealed that the alkylphenol ethoxylates gave poor results compared with the lower molecular weight alcohol ethoxylates. The tests revealed that in concentrated use a synergism did exist between the alkyl phenol ethoxylate and butyl cellosolve but that under high dilution conditions the solvent was ineffective. These findings are illustrated in the following diagrams.

Figure 7

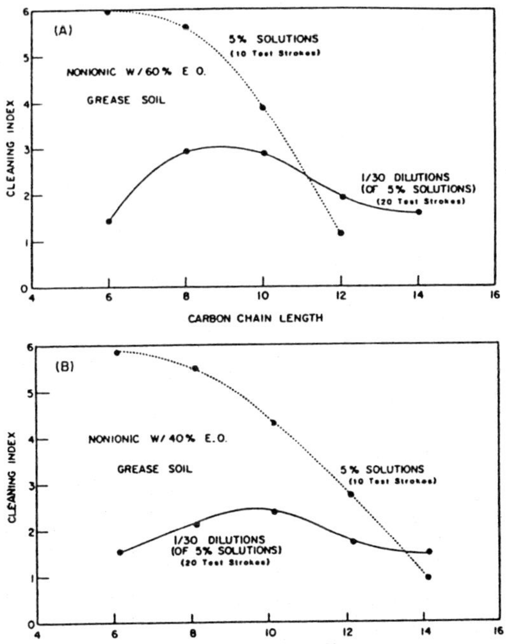

Effect of dilution on the hard-surface cleaning performance of nonionic homologs containing: (a) 60% ethylene oxide; (b) 40% ethylene oxide (using grease soil).

Figure 8

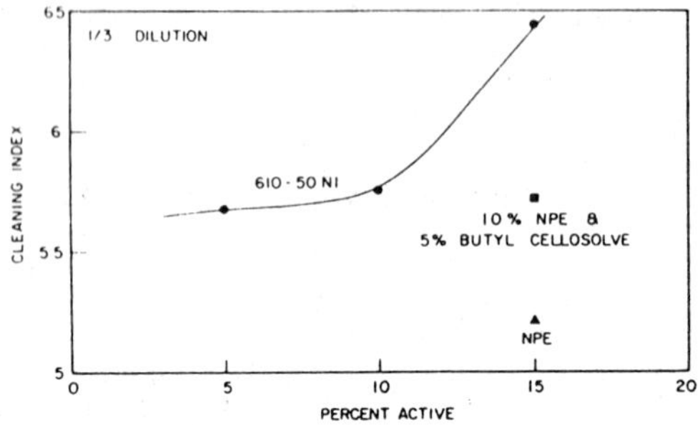

Comparison of the hard-surface cleaning performance of ALFONIC® 610-50 nonionic, nonylphenol ethoxylate (9½ mole) and a nonylphenol ethoxylate/butyl cellosolve mixture (using 1/3 dilutions of built formulations on grease soil).

I would like to close with some general remarks on laboratory evaluation procedures. The need is for reproducibility, accuracy and relevance to the problem and this requires standard soils, soiling methods, soil aging as well as appropriate substrates and soiling methods. This work together with stability studies as discussed earlier must serve only as a precursor to plant scale trials; it will serve only as an indicator that success is likely but it is vital before full scale trials can be undertaken with any confidence.

References

1. G.M.Gantz "Foaming" in Surfactant Science Series Vol. 1 Ed. by M.J.Schick, pub. Dekker 1976.

2. W.M.Mizuno "Dishwashing" in Detergency Theory and Test Methods Part III of Volume 5 in the Surfactant Science Series Ed. W.G.Collier and R.C.Davis, pub. Dekker 1981.

3. H.Schott "Removal of Particulate Soil" p.191 in Detergency Theory and Test Methods, Part 1 of Volume 5 in the Surfactant Science Series E. W.G.Cutler and R.C.Davis pub. Dekker 1972.

4. Ross J & Miles G.D.Oil Soap 18, 99 (1941)

5. S.Sugden, J.C.S., 125, 27, 1924

6. S.Sugden, J.C.S., 121, 858, 1922

7. T.Matson & M.Berretz Soap, Cosm. & Chem. Spec. 41, Feb. 1980

8. M.F.Cox & T.P.Matson JAOCS 60, 1170, 1983

9. United States Testing Co., Inc., 1415 Park Avenue, Hoboken, N.J., U.S.A.

10. (FCRA) Fabric Care Research Association Ltd., Forest House Laboratories, Knaresborough Road, Harrogate, North Yorkshire, HG2 7LZ

11. (EMPA) : Eidgenössische Material prüfungs - und Versuchsanstalt Unterstrasse 11, 9001 St.Gallen, P.O.B. 977 Schweiz.

12. L.E.Cohen & J.A.Hook, Plat and Surf. Fin 70, (March 1985)

13. M.F.Cox & T.P.Matson J.A.O.C.S. 61, 1273, 1984

14. American National Standard ANSI/ASTM D3556-79, 1979
Annual Book of ASTM Standards Part 30, 567, 1979

Coal–Water Slurries

By G. Bognolo
ICI EUROPA LIMITED, SPECIALITY CHEMICALS GROUP, EVERSLAAN 45, B-3078 KORTENBERG, BELGIUM

During the last decade the price escalation of hydrocarbon fuels has been a major driving force to focus the attention of industry and researchers on energy related matters. It has equally been a powerful stimulus to the identification of creative solutions for better energy utilisation as well as for the development of alternative sources of fuels.

It is not surprising that coal has been the centre of most of the investigations as it offers a number of unique features and objective advantages in comparison to hybrocarbons :

- it is available in virtually unlimited quantities
- it has a lower cost per unit energy output, at least in the economic scenario that has developed over the recent years
- it has a more balanced geographical distribution across the world

Coal, however, also presents a number of distinct disadvantages, notably :

- it causes considerable environmental problems, especially because of sulphur oxides emission and ash disposal. If this drawback can be substantially minimised through benefication processes, it nevertheless requires additional costly steps, among which the drying of the wet powder material to a moisture content suitable for combustion is particularly expensive and cumbersome.

- its solid physical form makes the handling, transport and storage more complex and expensive than for hydrocarbons. Coal as such cannot be moved through pipelines and, therefore, requires more costly forms of surface transport. Also in areas of difficult access, pipelines often represent the only envisageable form of transport partly because of constraints from the environment but, what is more important, because pipelines require much lower investment and maintenance costs if new lines of communication have to be established.
  Even when traditional forms of surface transport are acceptable or necessary, the transport efficiency as determined by the unit carrier volume needed to accommodate a unit weight of coal is at a clear disadvantage compared to liquid fuels.
  Last, at the level of power stations or industrial installations, coal requires large storage areas and the transfer to the combustion units is manpower and equipment intensive.

Because of these disadvantages, and because originally crude oil was economically competitive to coal, liquid fuels have become the major source of energy in industrialised countries with the consequence that over the last twenty or thirty years the majority of the newly built combustion units, or the restructured older ones, have been designed to burn liquids with low ash residues and atmospheric emissions.

It is this situation that has represented a major obstacle to a rapid and substantial reconversion to coal once the rising crude oil prices have put it at an economic disadvantage against coal.

However, there is no doubt that if coal could be "fluidised" and its environmental characteristics improved, then it would rapidly become a widely utilised fuel.

Coal can be fluidised through liquification or gasification processes, but these require comparatively sophisticated engineering and demand substantial capital investment, maintenance and operating costs.

'Coal water slurries' is the other alternative and the one that presents a number of significant advantages from the conceptual, economical and practical point of view.

Before examining such advantages in detail, let us briefly review the key features of coal water slurries.

As the name in itself suggest, coal water slurries are dispersions of coal in a water carrier. Basically two systems can be devised.

A. Diluted (40 - 50% w/w) slurries with coal particle sizes ranging from 0 to 0.5 millimetres.

This system is only of interest for pipeline transport in a turbulent flow situation. The coal is kept in suspension through the mechanical flow of the water and has such a particle size and distribution that sedimentation, inevitably occurring when the flow is interrupted, does not lead to sediments that are hard to disperse once the flow is reestablished again. It should be emphasised that because of this the coal particles must not be in a thermodynamically stable state of dispersion which, upon particles settling, would lead to the notorious "claying" phenomenon and the associated sediment dilatancy. Thus, for a turbulent flow situation the use of chemical dispersing agents must be avoided.

As mentioned before, diluted slurries are of interest only as a relatively cheap means of coal transportation. At the receiving station the slurry must be dewatered to low moisture content before combustion. This is a lengthy and expensive step that can present major problems in certain climatic conditions, for example during the winters of Central Siberia or Western Canada.

It is this system, to which I shall refer hereafter as CWS (although the name of CWF - 'coal water fuel' - might perhaps be more appropriate), that is of major interest and that can overcome the inherent coal drawbacks I have outlined before.

CWS can be produced with relatively simple and inexpensive milling equipment, and provided the appropriate chemicals (especially dispersants and dispersion stabilisers) are used in the process, it is possible to produce low-viscosity, non-settling slurries that can be transported, handled, stored and burnt like heavy fuel oils.

A wet milling of the coal is most effective to produce the optimum particle size for a high coal loading in the slurry. Thus the CWS preparation can be economically associated with a coal beneficiation step that, by reducing sulphur and ash levels, yields an environmentally acceptable fuel.

The equipment and maintenance costs for a CWS plant are negligible compared to a coal liquification unit and the making of the slurry (once the basic parameters for the mill and chemical dispersant have been fixed) is a straightforward process that can be run by unskilled operators.

Thus CWS offers a low investment and simple option for coal fluidification which makes it the ideal choice for exploiting coal resources in inaccessible areas where the installation of expensive and sophisticated plants for alternative fluidification processes would be problematic.

Another major advantage of CWS is that the slurry can be burnt as such and does not require de-watering or any other additional upgrading prior to combustion. Burner and boiler modifications can be kept to a minimum and boiler derating, usually in the region of 20%, is significant but quite acceptable.

A last but not least advantageous feature of CWS is that there are no limitations, in principle, to the type of coal to be used, although certain coal characteristics like porosity and surface oxidation may influence the minimum coal concentration achievable in the slurry. Even with particularly difficult coals, however, loadings of the order of 65% w/w can be achieved provided the appropriate dispersing agent is used. At this coal concentration the slurry can burn with a stable flame, notwithstanding relatively low volatile levels in the coal.

There is, however, a challenge that confronts every CWS producer, and this is the definition of the chemical system yielding slurries with properties responding to the specific requirements of the main, possible end-uses.

End-use	**Scenario**	**CWS requirement**
Electricity or heat generation for industrial installation, district heating, greenhouses	a) CWS production at a central location b) CWS distribution through surface transport c) storage and combustion at the end-user premises	a) coal loading around 70% w/w b) slurry stability for prolonged time in both static and dynamic conditions c) good combustion characteristics d) environmentally acceptable

End-use	Scenario	CWS requirement
Industrial intallations, power stations	a) CWS production at the coal production site b) CWS distribution through pipeline c) storage and combustion ar receiving station	a) coal loading sufficient to ensure good combustion; high loadings not necessarily sought c) good combustion characteristics d) environmentally accepted
Steel industry	blast furnaces fuel supplement	a) high coal loading b) dynamic stability; minimum storage stability
Coal liquification/gasification	coal slurry	a) high coal loading b) some dynamic stability; negligible storage stability
Fuel	diesel engine fuel	a) high coal loading; b) good rheology at low coal particle size and high coal loading c) trouble-free combustion d) ash-free slurry

The chemicals that may intervene in the preparation process of the slurry are summarised in Table 1. Their balanced utilisation enables the producer to obtain the desired characteristics for any specific application.

Table 1

CWS CHEMICALS

AGENT	FUNCTION	CHEMISTRY	RATE of USAGE kg/ton of coal based on 100 % active matter
Wetting agent	Ensure intimate contact between coal and water and displace air from the channels between and inside the coal agglomerates	Low molecular weight ethoxylated fatty alcohols	0 - 2
Dispersing agent	Assist in the comminution of coal in the wet milling process and maintain the particles formed in a dispersed state	Anionic (sulphonated) poly-electrolytes or nonionic (ethoxylated) dispersants	5 - 10
Dispersion stabiliser	Prevent the settling of the coal and the formation of dilatant sediments on storage	High molecular weight poly-ethylene glycol, polyacryl-amides, natually occurring polymers (eg. xanthan gum)	0 - 0.5
Foam suppressant	Prevent air entrainment and maintain low slurry viscosities in the coal milling process	Polysiloxanes, blends of fatty acid esters and salts	0.5 - 1.0
Biocide	Protect the other chemicals from biodegradation during slurry storage	Formaldehyde, cationics	0 - 0.1

Among the chemicals given in Table 1, the dispersing agent is certainly of major importance. Apart from the fact that in several instances it can also provide wetting and dispersion stabilistion, the dispersant is essential to obtain manageable viscosities at the high coal volume fractions needed for good combustion characteristics and acceptable economics.

Although an exceedingly large number of surface active agents is reported in patents and literature as functional dispersants for CWS, in practice it appears that anionic polyelectrolytes or ethylene oxide derivatives are favoured because of the large scale trials and semi-commercial operations already established.

The polyelectrolytes (lignin sulphonates or naphthalene sulphonates/formaldehyde condensates or polyacrylates) are attractive because of their low market price and large availability. Unfortunately they are effective (or cost-effective) only with certain types of coal, are not ideal for the combustion characteristics of the slurry and, if the neutralising cation is sodium, cause severe problems of slag formation in the bottom ashes.

The ethylene oxide derivatives also have an attractive cost/performance profile, can be made available in the volumes foreseen as necessary for the medium and long term future and can be tailored to provide optimum performance on different coal types. We have found that these products work on the broadest spectrum of coal types, do not give problems of slag formation and, most importantly, have no adverse effect on combustion.

A summary of the key features of anionic and nonionic dispersants is given in Table 2.

Table 2

CWS DISPERSANTS

Key features

	POLYELECTROLYTES	ETHOXYLATED DERIVATIVES
ADVANTAGES	Low additive cost Large availability	Excellent cost effectiveness Large availability No adverse effects on combustion Free from slag formation problems Broad spectrum of effectiveness Molecules can be tailor made to fit specific requirements
DISADVANTAGES	Effective only with certain coals Cost effectiveness sometimes dubious Not optimal for slurry combustian Slag formation problems (with sodium based products)	

Thus, on balance, dispersants based on ethylene oxide offer significant advantages and are gaining increasing acceptance over the better known and studied, but poorer performing, anionic dispersants.

An example of the dispersing effect achievable with anionic and non-ionic species is given in the following figure.

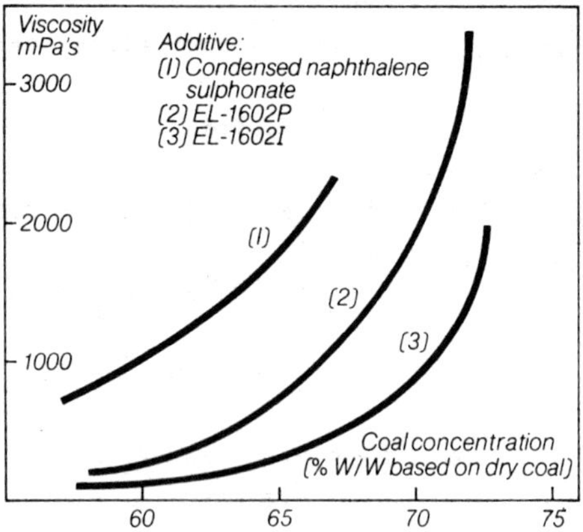

100 RPM Brookfield Viscosity of CWS's in function of Coal loading. Additive level: 1% w/w based on dry coal.

It is clear that the two nonionics ATLAS EL 1602 P and ATLAS EL 1602 I allow the production of slurries with higher solid loadings and better rheology than a plyelectrolyte, specifically recommended for CWS application.

As already mentioned before, nonionic dispersants can be designed to achieve maximum dispersion effectiveness with specific coals. The products work through steric stabilisation effects. They have a basic molecular structure consisting of an anchoring moiety that ensures the adsorption onto the dispersed phase surface and a stabilising moiety that provides stabilisation through osmotic and entropic effects.

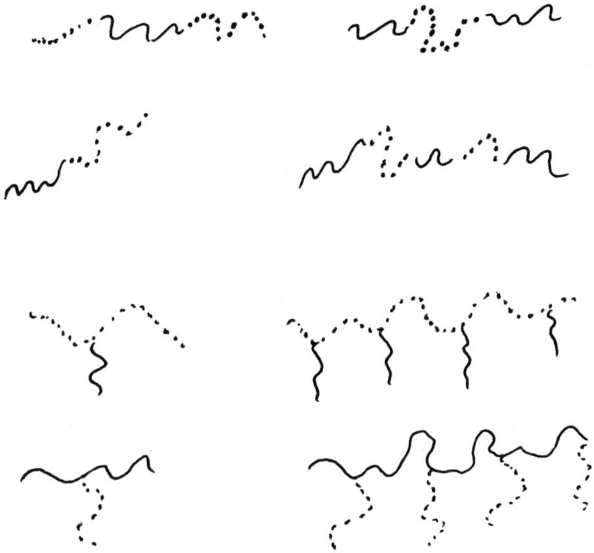

(________) stabilising moiety
(........) anchoring moiety

For optimum effectiveness the stabilising chains must interact strongly and be easily solvated by the aqueous phase. Yet the overall molecule solubility must not reduce the surface adsorption that results essentially from a phase rejection mechanism for the anchoring moiety.

The adsorption of the dispersant can be further enhanced by the inclusion of specific groups, but in this respect it should be noted that the surface of coal particles is extremely heterogeneous by nature and depends upon the coal type, history and treatment. Therefore, although it is possible to produce nonionic dispersants that are strongly adsorbed onto a very broad spectrum of coals, the success in producing a "universal", highly effective dispersing agent that can be used with any coal suspension still remains a challenge.

A review of CWS is not complete unless the economics and the future perspectives are discussed in some detail.

In a situation of a stable hydrocarbon fuel market, it would be both convenient and meaningful to analyse CWS economics on a relative basis using crude oil (or more accurately heavy fuel oil) average prices as reference.

However, in the market situation of today the reference point is subject to unforeseeable influences and fluctuation so much, that anything I could dare to write today might be contradicted by the time the paper will be delivered.

A few months ago there was a general consensus that CWS could be competitive to fuel oil if the crude oil prices were in the region of 25-28 US$/bbl. This was based on a total cost for the CWS of :

coal	:	2.1 US$/G.Joule
manufacturing	:	0.3 US$/G.Joule
chemicals	:	0.6 US$/G.Joule

and the following assumptions :

- CWS plant producing 1 million t/yr of slurry. The cost of such plant including coal storage facilities is in the region of 15 millin US$
- Return on investment in 2 years
- Coal concentration in the CWS : 70%
- Calorific value of the CWS : 20 M.Joule/kg
- Boiler derating in the CWS combustion plant : 20%

There is no doubt that at the spot prices of crude oil during the month of February 1986, CWS are not competitive on the basis of a simplified analysis that only takes into account straightforward economic and engeneering considerations like market prices, return on investments, combustion efficiencies.

This will remain true even if some of the cost assumptions outlined before should show (as they most surely will) a downward trend to adjust to the new economic realities created by reduced oil prices.

However, the scenario for CWS is much more complex and involves considerations and concepts exceeding and possibly outweighing the simple fuel replacement economics. Among these, at the top, rank :

- the need to develop a fuel that is more price stable, more widely available and of more secure supply than the hydrocarbon fuels, whilst retaining their advantages of conveniency for storage and handling

- through pipeline transport, CWS offer the possibility of using huge coal resources that would otherwise be inaccessible

- countries that have large coal resources and little crude oil might find CWS an attractive option to control their payment balance; countries that are rich in both coal and oil may still want to retain strategic reserve of a fuel that, on the basis of today's knowledge, will surely be depleted in a not too distant future.

Thus I believe it is not unrealstic to foresee a promising future for CWS and consequently substantial business opportunities for the surface active agents industry that will supply most of the chemicals used in their production.

This is further substantiated by the fact that the CWS technology has reached a stage where major projects can be initiated. It has been announced recently that, by 1987, the USSR will build and operate a 250 km pipeline to transport

CWS from the village of Belovo in the Kustzbas coal area to Novosibirsk. This project could involve transport of up to 3 million tons/year of coal and use about 30,000 tons/year of chemicals.

The USSR have an even more ambitious project under evaluation for pipeline transportation of CWS over 2,000 km and involving 10 times as much coal and chemicals as the Belovo project.

Other projects, both for demonstration purposes and actual CWS production, are underway in South Korea, Japan, Sweden, the Netherlands, Brazil, India, China and the U.S.A.

It is difficult, in the present situation, to attempt to quantify precisely what the CWS market could be over the next decade. The scale of any major CWS project is such, and the non-economic driving forces so fluctuating, that any estimate can easily carry a 100% error if not more.

I would guess, however, that by 1995 in a depressed crude oil price situation as the one of today, there could still be a worldwide market of around 15 million tons/year of CWS.

If one assumes an average loading of 700 kilos of coal per ton of slurry and an average use of chemicals (dispersants, antifoams, biocides, etc.) of 9 kilos per ton of coal, this correpsonds to about 100,000 tons/year of chemicals of which up to 100% and in any case no less that 80-85% would be surface active agents.

The market value could range between 150 and 200 million US$ per year.

If the crude oil prices should raise to the end '85 level, the market could probably double both in terms of tonnage and value.

At present the slurry and chemicals consumption are much more modest : I estimate that in 1985 a maximum of 300,000 tons of CWS (i.e. about 2,000 tons of chemicals) was used on a worldwide basis largely for semi-technical trials and the early commercial applications.

However, the state of the CWS technology, its features and advantages, the interest it continues to attract, even in an apparently unfavourable economic situation and the realisations underway, make it realistic to say that there is a future for CWS and that it will represent a large and profitable market for surface active agents in the decades to come.

Cationic Surfactants

By A. D. James, P.H. Ogden,* and J. M. Wates
AKZO CHEMIE UK LIMITED, HOLLINGWORTH ROAD, LITTLEBOROUGH, LANCS. OL15 OBA, UK

A cationic surfactant is an amphipathic species in which the hydrophilic head group carries a positive charge. Physico-chemical properties associated with respect to adsorption or chemsorption onto specific substrates cause this class of compounds to be of considerable commercial importance. Worldwide consumption is estimated to be in excess of 250,000 tons per annum, which corresponds to 5-10% of the total surfactant usage.

In principle the positive charge may be associated with various chemical functional groups and sulphoxonium (1), sulphonium (2), phosphonium (3) and iodonium (4) compounds are described in the literature as possessing useful bacteriostatic or insecticidal properties[1].

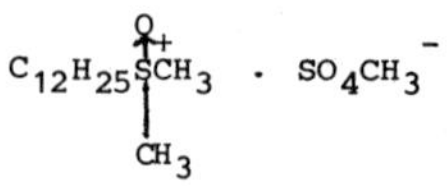

(1)

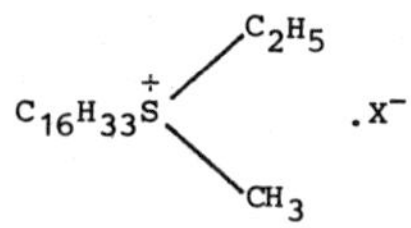

(2)

$$C_{16}H_{33}\overset{+}{P}(C_2H_5)_3 . Br^-$$

(3)

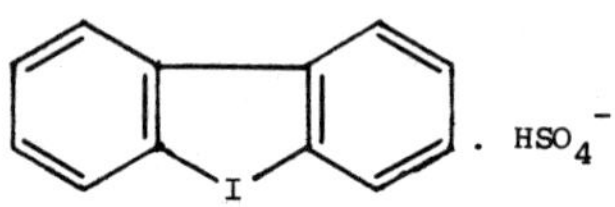

(4)

In practice only compounds in which the positive charge resides on a nitrogen atom contained in either a protonated amino (5) or a quaternary ammonium (6) compound are of commercial significance. Other materials are either too toxic or too expensive.

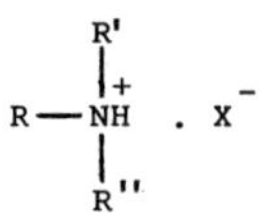

(5)

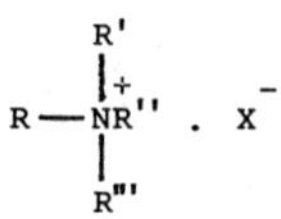

(6)

NITROGEN-CONTAINING CATIONIC SURFACTANTS

A wide variety of chemical compounds may contain protonated or quaternised amino groups which may be present as a constituent of a heterocycle such as imidazoline (7), morpholine (8) or pyridine (9). The alkyl group R is a hydrocarbon which may be linear or branched, saturated or unsaturated, complex terpenoid as in dehydroabietyl amine (10) or contain functional groups such as amido, ester, ether or amino.

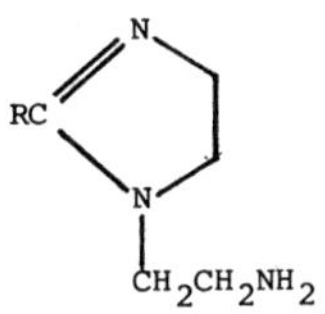

(7)

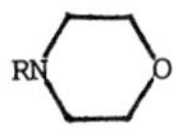

(8)

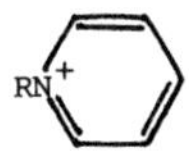

(9)

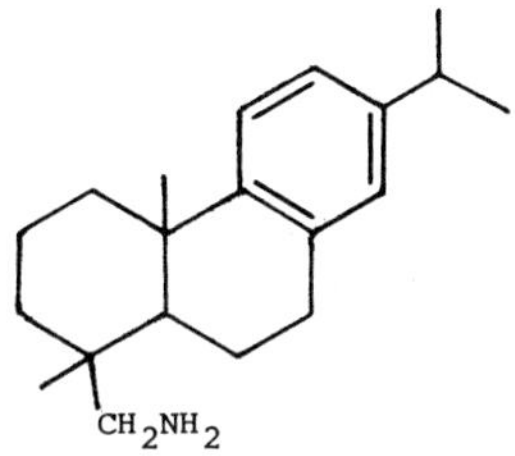

(10)

However, simple alkylamines in which R is a linear hydrocarbon and their derivatives (11-16) account for approximately 80% of the total cationic surfactant market.

$R{-}NH_2$ (11) R_2NH (12) R_3N (13)

RR'_2N (14) $RR'_3\overset{+}{N} \cdot X^-$ (15) $RR'_2R''\overset{+}{N} \cdot X^-$ (16)

HYDROPHOBIC GROUP

Most of the commercially available aliphatic amines are derived from naturally occurring fats and oils such as tallow, coconut, rapeseed and soya. Their alkyl chains are linear and contain the alkyl group distribution shown in Table 1. The degree of unsaturation depends upon the nature of the base oil from which they are derived and the choice of manufacturing conditions. Petroleum derived amines are also commercially available, in which case the alkyl chains may contain an odd or even number of carbon atoms, and may also be branched[2].

TABLE 1

APPROXIMATE COMPOSITION OF NATURALLY OCCURRING FATS AND OILS[3]

NATURAL PRODUCT	C_8	C_{10}	C_{12}	C_{14}	C_{16}	C_{18}	C_{20}	% UNSAT.
COCONUT	9	7	46	18	9	11		10
PALM KERNEL	3	7	47	14	9	20		19
TALLOW				6	27	66		52
LARD				1	31	66		56
PALM				2	32	67		61
SOYA					10	89	1	87
OLIVE				7	2	89		89
RAPESEED					1	47	51	98
HERRING				7	18	21	53	80
COD LIVER				6	28	30	35	85

SYNTHESIS OF ALIPHATIC AMINES

Synthesis of aliphatic amines from the corresponding carboxylic acid involves a two-step process. Aliphatic nitrile is obtained by dehydration of the intermediate ammonium carboxylate, usually through a continuous process in which ammonia is contacted with acid in the presence of catalyst under conditions which drive the equilibrium series towards nitrile and water (I)[4].

$$RCOOH \underset{-NH_3}{\overset{+NH_3}{\rightleftharpoons}} RCOONH_4 \underset{+H_2O}{\overset{-H_2O}{\rightleftharpoons}} RCONH_2 \underset{+H_2O}{\overset{-H_2O}{\rightleftharpoons}} RC{\equiv}N \qquad (I)$$

The resulting nitrile is then hydrogenated in the presence of catalyst to produce primary (II), secondary (III) or even tertiary (IV) amine depending upon the prevailing reaction conditions. High pressure and the presence of excess ammonia favour (II), while lower pressure and the continuous removal of ammonia favour (III) and (IV)[5].

$$RC{\equiv}N \xrightarrow{+H_2} RCH{=}NH \xrightarrow{+H_2} RCH_2NH_2 \qquad (II)$$

$$RCH_2NH_2 + RCH{=}NH \xrightarrow{+H_2} (RCH_2)_2NH + NH_3 \qquad (III)$$

$$(RCH_2)_2NH + RCH{=}NH \xrightarrow{+H_2} (RCH_2)_3N + NH_3 \qquad (IV)$$

Primary amine obtained by this process is usually at least 95% pure and this can be improved to 98% by distillation. The higher molecular weight secondary amine is usually at least 90% pure, always containing primary and/or tertiary amine. Although further purification of secondary could be achieved by distillation those of most commercial importance are derived from C_{12} to C_{18} acids and are too high boiling for this to be economic.

Secondary and tertiary amines prepared by the above process are symmetrical. Unsymmetrical secondary amines have been prepared by catalytic reduction of mixed nitriles or of a nitrile in the presence of an amine (V)[6]. This always produces mixtures however and unsymmetrical secondary amines have not attained commercial significance.

$$RC\equiv N + R'NH_2 \xrightarrow{(H)} RNHR' + NH_3 \qquad (V)$$

Unsymmetrical amines of the type RNR'_2 are prepared from the corresponding primary or secondary amine by reaction with alcohols (VI), alkyl halides (VII) and aldehydes (VIII).

$$ROH + R'_2NH \xrightarrow{H_2,cat} RNR'_2 + H_2O \qquad (VI)^7$$

$$RCl + R'_2NH \longrightarrow RNR'_2.HCl \qquad (VII)^8$$

$$R_2NH + CH_2{=}O \xrightarrow{H_2} R_2NCH_3 \qquad (VIII)^{9,10}$$

SYNTHESIS OF BRIDGED AMINES

The term 'bridged amine' has been used to describe materials in which the amino group and surfactant alkyl group are linked by another functional group. Typical of such are ether amines, prepared by cyanoethylation of an alcohol followed by hydrogenation (IX)[11], amido-amines from diamines and carboxylic acids (X)[12], amino-esters from ethanolamines and carboxylic acids (XI)[13], and imidazolines from cyclisation of certain amido-amines (XII).

$$ROH + CH_2{=}CHC\equiv N + H_2 \longrightarrow ROCH_2CH_2CH_2NH_2 \qquad (IX)$$

$$RCOOH + NH_2CH_2CH_2CH_2N(CH_3)_2 \longrightarrow RCONHCH_2CH_2CH_2N(CH_3)_3 \qquad (X)$$

$$[(CH_3)_2NH + CH_2{=}CHC\equiv N \xrightarrow{(H)} NH_2CH_2CH_2CH_2N(CH_3)_2]$$

$$RCOOH + (CH_2)_2NCH_2CH_2OH \longrightarrow RCOOCH_2CH_2N(CH_3)_2 \qquad (XI)$$

$$RCOOH + NH_2CH_2CH_2NHCH_2CH_2NH_2 \longrightarrow RCONHCH_2CH_2NHCH_2CH_2NH_2$$

$$\xrightarrow{-H_2O} R\underline{C{=}NCH_2CH_2N}CH_2CH_2NH_2 \qquad \text{(XII)}$$

Bridged amines are usually more conveniently prepared than are conventional amines since their synthesis does not require such high pressures. Occasionally the linking group can give rise to favourable physical properties as exemplified by the lower melting points of ether amines. However, it is frequently a point of potential chemical degradation, particularly with respect to hydrolysis.

CHEMICAL REACTIONS OF ALIPHATIC AMINES

Due to its lone pair of electrons the nitrogen atom is strongly nucleophilic and basic; basicity is reported to be in the order secondary > primary > tertiary. Amines react strongly with inorganic and organic acids to yield the corresponding amine salt (XIII). Amine hydrochlorides and low molecular weight acid salts are fairly soluble in water whereas sulphates, phosphates and salts of higher molecular weight acids are much less soluble.

$$RNH_2 + HA \longrightarrow RNH_3^+A^- \rightleftharpoons RNH_3^+ + A^- \qquad \text{(XIII)}$$

The amines undergo typical nucleophilic reactions such as alkylation, addition to activated double bonds and ring cleaving, and these reactions are of commercial significance since they can lead to the formation of specialty cationic surfactants as follows :

Alkylation

$$RNH_2 + R'X \longrightarrow RN^+R'_3.X^-$$

Converts primary, secondary or tertiary amine into a quaternary ammonium compound. Quaternaries are strongly ionised over a wide pH range. They are powerful surfactants and adsorb strongly onto negatively charged substrates over the full pH range.

Alkoxylation

$$RNH_2 + (x+y)\ \underset{\text{(ring)}}{CH_2-O-CH_2} \longrightarrow RN\begin{cases}(CH_2CH_2O)_xH\\(CH_2CH_2O)_yH\end{cases}$$

Converts a primary or secondary amine into an alkoxylated tertiary amine. Physical properties are controlled by the degree of alkoxylation. Low levels of ethoxylation, for instance, cause a decrease in water solubility and melting point. Thereafter water solubility and melting point increase with increasing ethoxylation. Under alkaline conditions these materials tend to behave like nonionic surfactants since the tertiary amine is not protonated. Under acid conditions they are cationic. Alkoxylated amines can be alkylated to produce quaternary ammonium salts.

Cyanoethylation, carboxymethylation and similar chemical synthesis techniques can be used to convert amines into amphoterics, typical of which are alkylamino acids, alkylbetaines and alkylsulphobetaines.

$$RNH_2 + CH_2{=}CHCOOH \longrightarrow RNHCH_2CH_2COOH$$

These materials can behave as cationic surfactants at low pH and anionic at high pH.

The above are the most commercially significant chemical modification procedures; allied to the wide range of available amines i.e. primary, secondary and tertiary from various feedstocks, they give enormous scope for the preparation of specialty chemicals with regard to such criteria as solubility, melting point, HLB value, optimum pH range etc.

Some other chemical modifications are summarised in fig. 1.

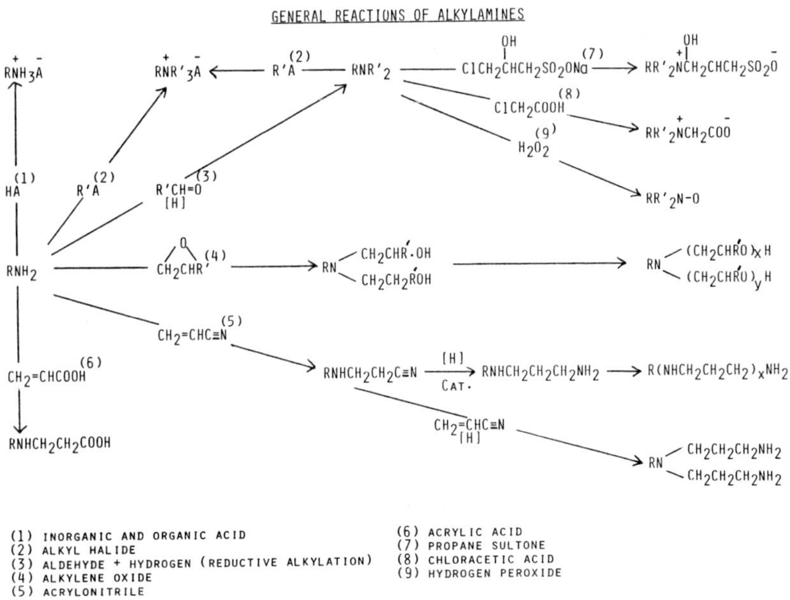

Fig. 1

MELTING POINTS OF ALIPHATIC AMINES

At higher molecular weights melting points tend towards that of the parent alkane, primary and secondary amines tending towards the corresponding linear alkane. Factors which affect melting point such as hydrogen bonding, increased ionic character, branching and unsaturation are most significant at lower molecular weights (fig. 2 and Table 2).

MELTING POINTS OF ALIPHATIC AMINES

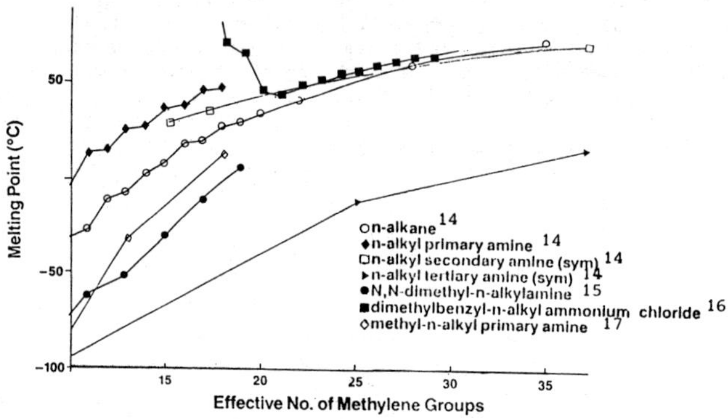

Fig. 2

TABLE 2

FACTORS AFFECTING MELTING POINTS OF AMINES

UNSATURATION		
Octadecylamine	$C_{18}H_{37}NH_2$	49°C [14]
Oleylamine	$C_{18}H_{35}NH_2$	15°C [18]
BRIDGING GROUPS		
Octadecylamine	$C_{18}H_{37}NH_2$	49°C
4-oxa-octadecylamine	$C_{14}H_{29}OC_3H_6NH_2$	28°C [19]
N',N'-dimethylheneicosamine	$C_{21}H_{43}N(CH_3)_2$	20°C [20]
N(N',N'-dimethylaminopropyl)	$C_{16}H_{33}CONHC_3H_6N(CH_3)_2$	60°C [21]

BRANCHING

No of carbons in akyl group	Melting Point of primary amine °C: Linear	α methyl [22]
10	7	-80
12	28	-29
18	49	10

SOLUBILITY OF AMINES

Water solubility of aliphatic amines of surfactant size is very low. Reaction with mineral acids or low molecular weight carboxylic acids produces the soluble amine acid salt. One protonated amino group will solubilise a twelve-methylene hydrocarbon chain. Two protonated amino groups or the more hydrophilic trimethyl ammonium group will solubilise an eighteen-methylene hydrocarbon.

Structural changes which have been shown to increase the water solubility of hydrocarbons such as branching or unsaturation also increase the water solubility of amines (Table 3). The low solubility of $C_{16}H_{33}N(CH_3)_2$ Benzyl chloride at 25 deg C compared with $C_{16}H_{33}N(CH_3)_3$ chloride at 40 deg C is thought to illustrate the Krafft effect, i.e. solubility of surfactant insufficiently high to reach C.M.C. at lower temperature.

TABLE 3

WATER SOLUBILITY OF ALIPHATIC AMINES

FREE AMINES

INSOLUBLE AT SURFACTANT MOLECULAR WEIGHTS

AMINE SALTS

SOLUBILITY DECREASES WITH INCREASING MOLECULAR WEIGHT AND LINEARITY

$C_{18}H_{37}NH_3$ ACETATE	1%
$C_{12}H_{25}NH_3$ ACETATE	30%
ROSIN AMINE ACETATE	50% [23]

DIAMINE SALTS

$C_{18}H_{37}NH_2C_3H_6NH_3$ DI ACETATE	5%
DI OLEATE	INSOLUBLE

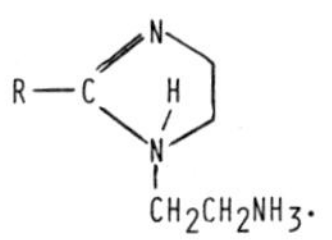

DI TRIMER ACID	INSOLUBLE IN WATER AND HYDROCARBON

QUATERNARY AMMONIUM

$C_{12}H_{25}N(CH_3)_3 \cdot Cl$	68% [24]	$C_{12}H_{25}N(CH_3)_2Bz$,	60% [16]
$C_{14}H_{25}N(CH_3)_3 \cdot Cl$,	61%	$C_{14}H_{25}N(CH_3)_2Bz$,	27%
$C_{16}H_{25}N(CH_3)_3 \cdot Cl$,	53%	$C_{16}H_{25}N(CH_3)_2Bz$,	0.85%
$C_{18}H_{25}N(CH_3)_3 \cdot Cl$,	42	$C_{18}H_{25}N(CH_3)_2Bz$,	0.1%

In non-polar organic solvents amines do not exhibit the critical micelle concentration effect[25,26]. They form oligomers due to stepwise aggregation of molecules with the maximum number of molecules per oligomer reported to be about ten, and optical density measurements of iodine in toluene in the presence of tetrakis dodecylammonium bromide indicate a smooth relationship between concentration and optical density (fig. 3).

Solubilization of iodine in toluene

by the presence of $(C_{12}H_{25})_4NBr$ [27]

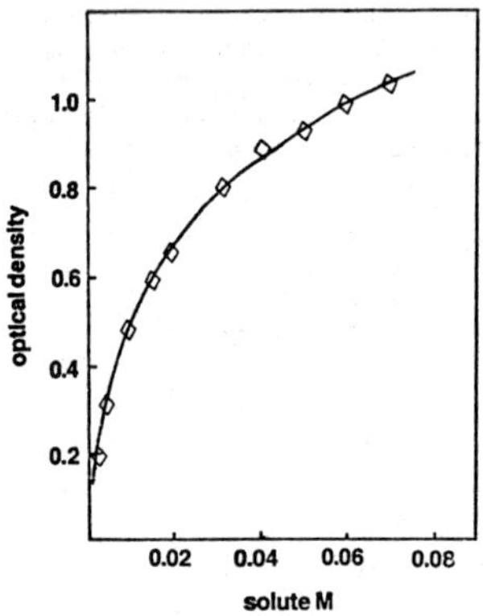

Fig. 3

In polar solvents such as dimethyl formamide or dimethylsulphoxide a critical micelle concentration has been reported[28].

In mixed aqueous solvent solubility has been reported to decrease with increasing molecular weight and ionic character (Table 4).

TABLE 4

SOLUBILITY OF AMINE ACETATES AND HYDROCHLORIDES IN 95% AQUEOUS ETHANOL[29,30]

ALKYL GROUP	% DISSOLVED AT 20°C AND 40°C			
	ACETATE		HYDROCHLORIDE	
	20	40	20	40
C12	45	72	15	46
C14	29	61	9	28
C16	22	48	4	14
C18	0.5	32	2	6

SURFACTANT BEHAVIOUR

Like anionic surfactants, cationics exhibit effective rather than efficient behaviour due to their charged hydrophile which disfavours micelle formation at low concentrations[31], and they exhibit higher C.M.C. values than those of corresponding nonionic surfactants (Table 5).

TABLE 5

CRITICAL MICELLE CONCENTRATIONS OF ANIONIC, CATIONIC AND NONIONIC SURFACTANTS[32]

SURFACTANT		Alkyl Group (R)			
NONIONICS		C_{18}	C_{16}	C_{14}	C_{12}
	$R(OCH_2CH_2)_6OH$	$3x10^{-7}$	$9x10^{-6}$	$1x10^{-5}$	$8x10^{-5}$
	$R(OCH_2CH_2)_9OH$	$5x10^{-7}$	$2x10^{-6}$		$9x10^{-5}$
ANIONICS					
	$ROSO_3Na$	$2x10^{-4}$	$5x10^{-4}$	$3x10^{-3}$	$1x10^{-2}$
	RCOONa	$1x10^{-3}$	$3x10^{-3}$	$7x10^{-3}$	$2x10^{-2}$
CATIONICS					
	RNH_3Cl	$3x10^{-4}$	$.9x10^{-3}$	$3.6x10^{-3}$	$1.6x10^{-2}$
	$RN(CH_3)_3Cl$	$4x10^{-4}$	$1.4x10^{-3}$	$3.7x10^{-2}$	$1.8x10^{-2}$
	$(C_8H_{17})_2\overset{+}{N}(CH_3)_2Cl^-$		$(2.66x10^{-2})$		

Branching, which has been shown to increase effective behaviour in anionics (fig. 4), also increases effectiveness of cationics whether the branching is at the nitrogen atom (fig. 5), or in the alkyl chain[33].

Surface tension of aqueous solutions of sodium p-dodecylbenzenesulfonates at 75°C as a function of their concentration[31]

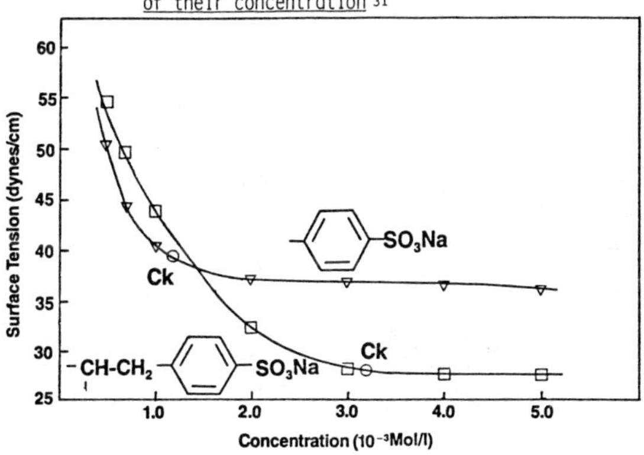

Fig. 4

Surface tension of aqueous solutions of dialkyl quaternary ammonium compounds at 20°C as a function of their concentration[34]

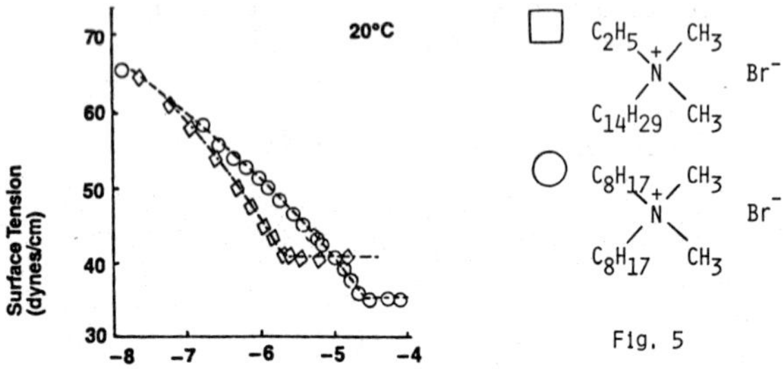

Fig. 5

HYDROPHILE-LIPOPHILE BALANCE

The classical HLB value concept which was developed for nonionic surfactants does not appear to hold well for charged surfactants. Modified expressions have been proposed[35,36], but these are reported to be erroneous when applied to ethoxylated amines.

The main application of the HLB concept is the assessment of emulsifier performance; since most industrial applications of cationic surfactants involve surface modification there are few reported HLB values.

The partition coefficient of cationic surfactants between different solvents is probably a more valuable factor since this can give an indication of cost efficiency. For example the partition of corrosion inhibitor between crude oil and produced water in pipeline protection is a prime consideration when selecting surfactants for this application.

There are few references to the determination of partition coefficients of cationic surfactants. Data presented in fig. 6 describes n-alkyl primary amines and n-alkyldimethylbenzyl ammonium bromides in water/octanol[37], and n-alkyl primary amines[38] and n-alkyl carboxylic acids[39] in water/heptane.

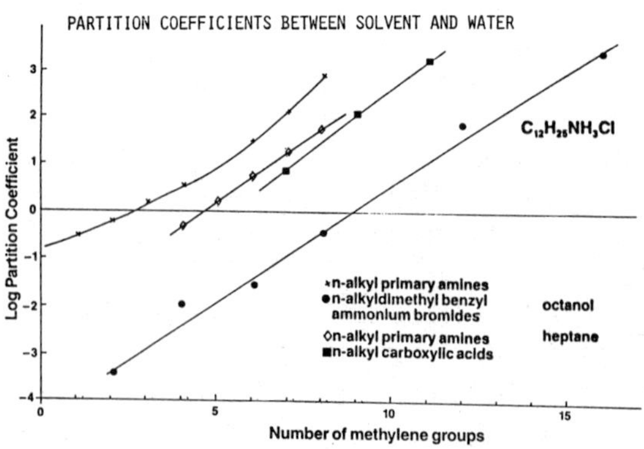

Fig. 6

SURFACE CHARGE

In the presence of water the surfaces of many substances adopt an electric charge, the polarity of which is dependent upon the prevailing pH, the acidity of the surface groups and other potential determining ions present in the solution.

For example, hydrolysis of the Si-O-Si bonds at the surface of silicates can produce Si^+ or SiO^- through the following reactions (XIIII).

$$\begin{array}{lll} Si - O - Si & \longrightarrow & Si^+ + SiO^- \\ Si^+ + H_2O & \longrightarrow & SiOH + H^+ \\ SiOH + H^+ & \longrightarrow & SiO^- + 2H^+ \end{array} \qquad \text{(XIIII)}$$

Positively charged cationic surfactants are strongly adsorbed on surfaces at pH values higher than the zero charge pH, typical values of which are as follows[40,41]; Muscovite, 0.95; Quartz, 1.8; Bentonite, 3.0; Kaolinite, 3.4; Eggonite, 4.0; Andalusite, 5.2; Cummingtonite, 5.2; Wool, 3.5; Polyamide, 5.2. As a result they are substantive to mammalian hair and skin, cotton, foliage, wood, paper, minerals, glass and many other surfaces under normal use conditions. In contrast, anionic surfactants are most strongly adsorbed below the zero point.

Although negative surface charge increases at high pH, the degree of protonation of the amino group decreases and for a given substrate there is an optimum pH corresponding to the maximum degree of adsorption of a specific amine

TOXICOLOGICAL ASPECTS

As with anionic surfactants, the lower molecular weight members of the cationic surfactant homologous series are quite toxic (acetic acid, methylamine, methanesulphonic acid). Due to increasing insolubility higher members of the series are absorbed into animal tissues at a much reduced rate and toxicity correspondingly decreases.

Lower molecular weight alkylamines i.e. C_{18} and less are classified corrosive, whereas higher molecular weight amines and most quaternaries are classified irritant. For comparison, sodium dodecylsulphate is classified harmful.

Toxicity is an important consideration when making a surfactant choice. Skin and eye irritation data is particularly important for quaternary ammonium compounds which find consumer application as disinfectants, deodorants and fabric conditioners. Rabbit eye irritation data (Table 6) indicates that aqueous solutions of quaternaries with more than twelve methylene units in the alkyl chain have low irritancy.

Acute toxicity of alkylamines is limited by their low water solubility but modifications which increase solubility, such as ethoxylation, cause increased toxicity. Thus LD_{50} tallow alkylamine acetate > 2500 mg/kg, LD_{50} tallow alkylamine pentaethoxylate 5-600 mg/kg. Increased ethoxylation causes decreased toxicity but this is probably due to a dilution effect (Table 7).

TABLE 6

TOXICITY OF CATIONIC SURFACTANTS[42]

EYE IRRITANCY RABBIT
0.1 ML OF (A) 1% AND (B) 10% SOLUTION
MAXIMUM POSSIBLE SCORE 100
CRITICAL SCORE 30

	1% SOLUTION	10% SOLUTION
$RN(CH_3)_3Cl$		
R = 12	3.6	59.6
16	3.6	47.5
18	-	11.9
$R_2N(CH_3)_2Cl$		
R = HYDROGENATED TALLOW	-	11.7
TALLOW - $NH_3 \cdot CH_3COO$	15.9	33.3
TALLOW - N $\langle$ $(CH_2CH_2O)_xH$, $(CH_2CH_2O)_yH$		
X + Y = 5	-	18.8
50	-	0

TABLE 7

ACUTE TOXICITY OF CATIONIC SURFACTANTS LD_{50} (MG/KG)[43]

ALKYL GROUP	C_4	C_8	C_{10}	C_{12}	C_{14}	C_{16}	C_{18}	COCO	TALLOW	SOYA	DEHYDRO ABIETIC
PRIMARY AMINE	500^b	100^a	280^b 350^c	50^a 1020^b 1160^d		200^a	250^a				2500^b
PRIMARY AMINE ACETATE									2000^b		850^b
N,N DIMETHYL AMINE				1380^b		800^b 4290^c	780^b 4290^c	1580^b 4290^c		835^b 3000^c	
TRIMETHYL QUATERNARY (Cl OR Br)				250^b		250^b (Cl) 410^b (Br)		500^d	2600^d	800^b	
BENZALKONIUM CHLORIDE	495^d	790^d	660^d	700^d	1410^d	4800^d					
DIALKYL DIMETHYL QUATERNARY							1100^b	7000^b (HT)			
5 MOLES ETHOXYLATE									$5\text{-}600^b$		
50 MOLES ETHOXYLATE									$>2000^b$		
ALKYLBENZENE SULPHONATE, Na SALT				1260^b 2000^d							
ALKYL SULPHATE, Na SALT				1288^b							

a - ipr mouse b - oral-rat c - skin-rabbit d - oral-mouse

Amphoteric surfactants are often considered to be less toxic than other cationics because they are used in baby shampoos. This is slightly misleading. In the absence of other surfactants they can be severe irritants but they exhibit a mitigating effect upon anionic surfactants (Table 8).

TABLE 8

TOXICITY OF CATIONIC SURFACTANTS MITIGATING EFFECT OF AMPHOTERICS [44]

	COMPOUND	EYE IRRITANCY OF 15% AQUEOUS SOLUTION
A	SODIUM DODECYLBENZENE SULPHONATE 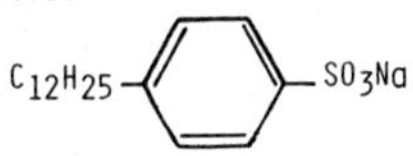	MODERATE
B	3-(DIMETHYLHEXADECYL AMMONIUM) PROPANE SULPHONATE $C_{16}H_{33}N(CH_3)_2CH_2CH_2CH_2SO_3^-$	SEVERE
	50% A + 50% B	NONE

CHOICE OF CATIONIC SURFACTANT

The wide choice of surfactant properties made available by chemical modification of the amino functional group gives this class of substance many industrial applications, some of which are illustrated in Table 9.

TABLE 9

INDUSTRIAL APPLICATIONS OF CATIONIC SURFACTANTS

ACID-THICKENING AGENT	ETHOXYLATED AMINES
ADHESION AGENT	DIAMINES
ANTICAKING AGENT	AMINE, AMINE SALT
ANTISTATIC AGENT	ETHOXYLATED AMINES
BIOCIDES	QUATERNARY, AMINE SALT
COSMETICS	QUATERNARIES, AMPHOTERICS
CORROSION INHIBITORS	AMINES. AMINE SALTS
EMULSIFYING AGENTS	ETHOXYLATED AMINES
FABRIC CONDITIONERS	QUATERNARIES
FUNGICIDES	DIAMINE CHELATES
GERMICIDES	QUATERNARIES
HERBICIDAL ADJUVANTS	ETHOXYLATED AMINES
MINERAL PROCESSING AGENTS	AMINE SALTS, QUATERNARIES
MOULD RELEASE AGENTS	AMINES
ORGANOPHILIC CLAYS	QUATERNARIES, AMINE SALTS
PIGMENT DISPERSANTS	QUATERNARIES, AMPHOTERICS
SPIN BATH ADDITIVES	ETHOXYLATED AMINES
STERILANTS	QUATERNARIES
SUGAR REFINING ADDITIVE	QUATERNARIES

Surfactant selection for most of these applications is dependent upon several factors. In addition to surfactant properties such as HLB value and CMC it is usually necessary to consider flammability, toxicity, biodegradability, and physical properties which determine the ease with which the material can be handled such as melting point, solubility or dispersability in solvents.

Final choice usually involves a compromise involving the above considerations, compatibility with the total system and cost.

The examples in Tables 10 and 11 illustrate how these factors may have dictated surfactant choice in two specific cases.

TABLE 10

RINSE APPLIED FABRIC CONDITIONER

PERFORMANCE REQUIRED	SURFACTANT CHOICE
Impart lubricant and antistatic properties to treated fabric.	
Strongly adsorbed onto fibre surface from aqueous medium at neutral or higher pH.	Quaternary ammonium compound
Performance effective, most or all of surfactant adsorbed rapidly onto fibre.	Low water solubility.
Low toxicity and skin irritation.	High molecular weight quaternary
Optimum performance.	$C_{18}<DiC_{12}<DiC_{18}<C_{22}$ [45]
Available in convenient retail form i.e. solution or aqueous dispersion.	DiC_{18} has lamellar structure and produces stable dispersions.[46]
Cost effective.	DiC_{18} ex tallow, C_{22} ex fish oil or rapeseed.
Most frequently used products	$(HT)_2\overset{+}{N}(CH_3)_2\overset{-}{Cl}$ 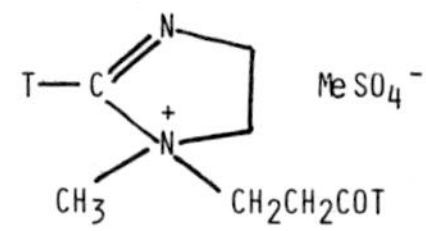

TABLE 11

BITUMEN EMULSIFIER

PERFORMANCE REQUIRED	SURFACTANT REQUIREMENTS
Emulsifier to produce bitumen-in-water emulsions.	Anionic, cationic, or nonionic surfactants with high HLB values.
Emulsion destabilised in presence of mineral aggregate due to strong adsorption of emulsifier from emulsion onto aggregate surface.	Cationic surfactant for most aggregates. Amine salt if application is allowable at low pH. C_{12} amine or C_{18} diamine give required HLB value.
Adsorbed emulsifier promotes strong bond between hydrophilic mineral surface and hydrophobic bitumen.	Diamines preferred due to improved adhesion properties.
Most frequently used surfactants:	
Rapid breaking emulsions	Tallowalkyl $NHCH_2CH_2CH_2NH_2$ as hydrochloride salt.
Slow breaking emulsions	Tallow $\overset{+}{N}(CH_3)_2CH_2CH_2CH_2\overset{+}{N}(CH_3)_3.2Cl^-$

FABRIC CONDITIONING

The largest single application of cationic surfactants is in the manufacture of rinse applied fabric conditioners. These products must be capable of being adsorbed from rinse water onto fabric, causing garments to dry more quickly, iron more easily, stay cleaner, feel softer and smell more pleasant than when left untreated. The following factors are important in influencing the selection of a suitable surfactant for use as a fabric conditioner.

During the wash cycle, which is usually conducted under alkaline conditions, fabric acquires a negative charge; thus surface modification of the type required may only be achieved using a cationic surfactant. Residual wash liquor which pollutes the rinse water raises its pH to a level at which only quaternary ammonium compounds retain their strongly substantive properties.

The need for rapid exhaustion of surfactant from the rinse water into the fabric demands high molecular weight quaternaries with low water solubility, as does the requirement that the surfactant should be non-irritating.

Performance testing of quaternary ammonium compounds which comply with the above constraints gave a panel rating order $C_{18} < diC_{12} < diC_{18} < C_{22}$ for the hydrophobic group and on these findings a C_{22} compound would appear to be preferable. However, one requirement of a retail product is that it can be distributed easily, and whereas a di C_{18} quaternary will form relatively concentrated aqueous dispersions due to the stabilisation of its lamellar structure by the presence of two long alkyl chains, a mono C_{22} quaternary will not.

Therefore, regardless of the cost difference between the two best performers, di C_{18} compounds are preferred due to their improved compatibility with the system; almost all fabric softener formulations are based on either di-hydrogenated tallow dimethyl ammonium chloride, or ditallow imidazolinium sulphate which are marketed as solutions in isopropanol.

Although the former is generally considered to give the better performance it is a low melting paste which must be processed in the molten state and dispersed in warm water during manufacture of the product. Smaller operators who are not equipped to handle pastes prefer to use the tallow based imidazoline which is liquid at ambient temperature and may be dispersed in cold water.

BITUMEN EMULSIFICATION

Bitumen has long been used as a binder to hold aggregate into the wearing surface of roads. It is applied to the road surface in a molten state, usually cut back with solvent to lower its viscosity, and chippings are susequently spread onto it and pressed into its surface by the use of a road roller. The resulting bond between the hydrophilic aggregate and the hydrophobic bitumen is not always good, particularly if the aggregate surface is damp. Improved wetting of the aggregate can be obtained by the addition of alkylamines to the bitumen; these are so substantive to many minerals that they displace surface water and render the mineral hydrophobic. Alkylamines used for this purpose are referred to as wetting or adhesion agents. The cationic surfactant most widely used as a wetting agent is the diamine derived from octadecylamine via cyanoethylation and hydrogenation.

Increasingly, the use of cut-back bitumen is being replaced by cationic bitumen emulsions, typically containing 60-70% bitumen. These are used in the same way as cut-back bitumen but are pumpable liquids at relatively low temperatures (around 90 deg C, compared with 140 deg C for cut-back bitumen). Since bitumen emulsions do not contain volatile fluxing solvents they have environmental advantages.

The mode of operation of bitumen emulsions depends on their destabilisation due to adsorption of the emulsifying agent onto the negatively charged aggregate, followed by breaking of the emulsion. Bitumen emulsions can be stabilised by most cationic surfactants provided their HLB values are correct, or for that matter by anionic or nonionic surfactants. However, the total system requires a surfactant which will be rapidly adsorbed onto the mineral surface and, having been adsorbed, will behave as a surface modifying agent, promoting strong adhesion between aggregate and binder. Diamines derived from tallow amine have been found to satisfy all these criteria better than other surfactants, either as the protonated amine for rapid breaking emulsions (which are used at low pH) or as quaternaries for use with slow setting emulsions.

REFERENCES

(1) Cationic Surfactants, Ed. E. Jungermann, Marcel Dekker, Inc., New York (1970) pp 191-199.

(2) Cationic Surfactants, Ed. E. Jungermann, Marcel Dekker, Inc., New York (1970) pp 148-178.

(3) A.E. Bailey, Industrial Oil and Fat Products, Interscience Publishers, New York (1945).

(4) R.H. Potts and C.W. Christensen, U.S. Pat. 2,314,894 (1943).

R.H. Potts, U.S. Pats. 2,414,393 (1947), 2,448,275 (1948), 2,504,045 (1950), 2,524,831 (1950).

R.H. Potts and R.S. Smith, U.S. Pat. 2,808,426 (1957).

R.H. Potts (Armour & Co.), U.S. Pat. 3,299,117 (1967).

(5) W. Schmidt (I.G. Farbenindustrie A-G), German Pat. 648,297 (1933).

(6) S.H. Shapiro (Armour & Co.), U.S. Pats. 2,781,399 (1957), 2,811,556 (1957).

(7) J. Westland, S.H. Shapiro and G. Michaels, French Pat. 1,104,778 (1964).

(8) C. Westphal and D. Jerchel, Ber., 73B, 1002 (1940).

(9) S.H. Shapiro and F. Pilch (Armour & Co.), U.S. Pat. 3,136,819 (1964).

(10) H.T. Clarke, H.B. Gillespie and S.Z. Weisshaus, J. Am. Chem. Soc., 55, 4571 (1933).

(11) J. Hood (American Cyanamid Co.), U.S. Pat. 2,371,892 (1945), E.L. Carpenter, U.S. Pat. 2,372,624 (1945).

(12) E.W. Cook and P.H. Moss (American Cyanamid Co.), U.S. Pat. 2,459,062 (1949).

(13) O. Albrecht and E. Matter (Ciba Ltd.), Swiss Pat. 334,628 (1959).

(14) C.R.C. Handbook of Chemistry and Physics, 60th ed. (1979).

(15) Akzo Chemie Prod. Bulletin.

(16) R.A. Cutler, E.B. Cimijotti, T.J. Okolowitch and W.F. Wetterau, Soap Chem. Spec., 43 (3), 84 (1967).

(17) Armour Ind. Chem. Co. Prod. Bulletin.

(18) Hoechst Technical Bulletin.

(19) Akzo Chemie Technical Bulletin.

(20) Estimated from fig. 2.

(21) N. Parris, J.K. Weil and W.M. Linfield, J.A.O.C.S., 50, 509 (1973).

(22) Armour Ind. Co. Tech. Bulletin (1968), Jungermann p162.

(23) Hercules Prod. Data Bulletin, O.R.132C Amine D.

(24) At 40°C; S.H. Shapiro, Fatty Acids and their Industrial Applications, Ed. E.S. Pattison, Marcel Dekker, Inc., New York (1968) pp109-119.

(25) A.S. Kertes, Micellization, Solubilization, Microemulsions [Proc. Int. Symp.] 1976 (Pub. 1977), 1, 445-454.

(26) E. Ruckenstein and R. Nagarajan, J. Phys. Chem., 84, 1349-1358 (1980).

(27) H. Gutmann and A.S. Kertes, J. Coll. Int. Sci., 51, (3), 406 (1975).

(28) H.N. Singh, S.M. Saleem, R.P. Singh and K.S. Birdi, J. Phys. Chem., 84, (17), 2191-4 (1980).

(29) R.S. Sedgwick, C.W. Hoerr and A.W. Ralston, J. Org. Chem., 10, 498 (1945).

(30) H.J. Harwood, A.W. Ralston and W.M. Selby, J. Am. Chem. Soc., 63, 1961 (1941).

(31) M.J. Rosen, J. Am. Oil Chem. Soc., 49, (5), 293-7 (1972).

(32) P. Mukerjee and K.J. Mysels, Critical Micelle Concentrations of Aqueous Surfactant Systems, NSRDS-NBS 36 (1971).

(33) P.A. Demchenko, I. Ya. Litvin, Yu. V. Tanchuk and B.S. Shapoval, Fiz.-Khim. Mekh. Liofil'nost Dispersnykh Sist., 4, 85-7 (1973).

(34) O. Dusart, Fluid Phase Equilib., 20, 265-8 (1985).

(35) J.T. Davies, Proc. Int. Congr. Surface Activity 2nd, London, 1, 426 (1957).

(36) I.J. Lin, J.P. Friend and Y. Zimmels, J. Coll. Int. Sci., 45, (2), 378-85 (1973).

(37) A. Leo, C. Hansch and D. Elkins, Chem. Revs., 71, (6), 525 (1971).

(38) J.M. Wates, Unpublished Results.

(39) R. Smith and C. Tanford, Proc. Nat. Acad. Sci., U.S.A., 70 (2), 289 (1973).

(40) Flotation, Ed. M.C. Fuerstenau, A.I.M.E., New York (1976), Vol. 1, p171.

(41) Cationic Surfactants, Ed. E. Jungermann, Marcel Dekker, Inc., New York (1970) p373.

(42) Armour Industrial Chemical Co., Products Data Bull. No. 67-14 (1967), Jungermann p591, Appendix 4.

(43) Registry of Toxic Effects of Chemical Substances, U.S. Dept. of Health and Human Services (1980).

(44) R. Ernst, Amphoteric Surfactants, Ed. B.R. Bluestein and C.L. Hilton, Marcel Dekker, Inc., New York (1982), p149.

(45) W.P. Evans, Chem. Ind., 893 (1969).

(46) A.D. James and P.H. Ogden, J.A.O.C.S., 56 (4), 542 (1979).

Cationic Surfactants in Mineral Processing

By H. L. Shergold
BP RESEARCH CENTRE, CHERTSEY ROAD, SUNBURY-ON-THAMES, MIDDLESEX TW16 7LN, UK

INTRODUCTION

Few of the metals or minerals required by modern civilisation are mined naturally pure and ready to use. Instead they occur in complex aggregates of up to twenty different minerals of which only one or two will be of value. There are some two hundred of these value minerals among the several thousands that go to form the earth's crust. The number of combinations of minerals that can occur with a valuable component is therefore very large and each deposit containing nominally the same valuable minerals will differ materially from all others. Consequently, the best separation process for a particular ore is closely related to its overall mineralogy and not to just the nature of the valuable components.

During the separation procedures the minerals are separated into distinct products by the exploitation of differences in chemical or physical properties of the various species. These properties may include size, shape, colour, specific gravity, electrical conductivity, magnetic susceptibility, surface chemical effects etc. Separation can only be achieved, however, if the valuable components are presented to the separating force as individual, unlocked or liberated grains. Liberation is usually achieved by crushing and grinding.

Flotation is one of the most important processes used to separate mineral particles and it is the main consumer of surfactants in the whole area of mineral processing. Over 10^9 tonnes of ore are treated by flotation annually and plants capable of treating over 100,000 tonnes of low grade material per day are not uncommon. Judged by the scale on which it is operated, it is not only a very important separation process but a very impressive technical application of surface chemistry and the use of surfactants. It is the main user of cationic surfactants in mineral processing. This paper will therefore be directed at the use of cationic surfactants in mineral flotation.

The success of the flotation process depends on rendering selected minerals in a pulp hydrophobic and hence floatable while keeping, or making, all the other minerals hydrophilic. Air bubbles blown into the pulp collect the hydrophobic particles and carry them to the top of the separation cell, where they concentrate in a froth which overflows into a launder. The hydrophilic minerals or tailings are recovered from near the bottom of the flotation cell. Typically, the cell is a stirred tank which might contain 15 - 30% by volume of entrained air with bubbles ranging mainly from 0.1 - 5 mm in diameter, the pulp density might be 25 - 40% by weight solids having a particle size 10 - 100 μm. The pulp is vigorously agitated to ensure that bubble-particle collisions occur. The froth at the top of the cell is generally produced by a non-ionic surfactant referred to as a frother.

Most minerals when crushed and ground in water are hydrophilic and selective hydrophobicity is obtained by adsorbing surfactants known as collectors at the mineral/water interface. These reagents eliminate or shield polar sites in the mineral surface which would otherwise hydrogen bond with or interact with water molecules. Collectors are characterized by (i) a polar group, usually ionic which interacts with the mineral surface and gives the reagent solubility, and (ii) a non-polar group, typically an alkyl chain, of sufficient size to impart hydrophobicity. They can be divided into two classes based on whether they are used to float non-sulphide or sulphide minerals. Typical collectors in the first class are long chain alkyl sulphates, sulphonates, carboxylates and amines, while short chain alkyl xanthates and dithiophosphates are typical thio-type collectors used in the flotation of sulphides. The main use of cationic surfactants in mineral processing is in the flotation of non-sulphide minerals.

Selection of the appropriate collector for a mineral separation process is dependent on the surface chemical properties of the minerals to be separated and the solution properties of the collector.

Solution Properties of Cationic Collectors

a) Ionization A considerable number of cationic surfactants have been either used or suggested as flotation collectors. Several such compounds are listed in Table 1. The amines, particularly primary amines, are the most widely used cationic collectors.

Table 1 Cationic Collectors

Collector	Structure
Primary amine salt	$RNH_3^+\ Cl^-$
Secondary amine salt	$R\ R^1NH_2^+\ Cl^-$
Tertiary amine salt	$R(R^1)_2\ NH^+\ Cl^-$
Quaternary ammonium salt	$R(R^1)_3\ N^+\ Cl^-$
Alkyl pyridinium salt	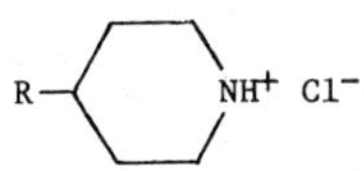
Hydroxyethyl-2-alkylimidazolines	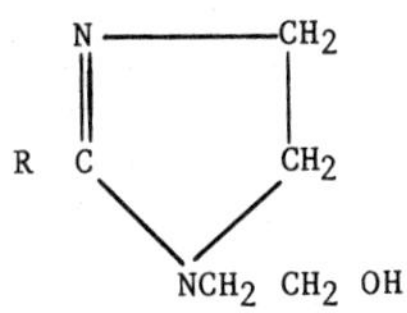
Sulphonium salt	$R\ S(R^1)_2^+$

R represents an alkyl chain, usual 10 to 20 carbons, and R^1 is a short alkyl chain usually a methyl group. Variants on the primary amines currently available as collectors include the ether amines with the general structure $R\text{-}O\text{-}CH_2.CH_2CH_2.NH_2$. Alkyloxydiamines are also available as collectors.

The collecting properties of these surfactants depend on such characteristics as ionization constant, solubility, ionic head group size, hydrocarbon chain length and critical micelle concentration (cmc). Primary, secondary and tertiary amines are weak bases, whereas quaternary amines are strong ones. Quaternary amines are completely ionized across the practical pH range whereas ionization of the other amines is pH dependent. Ionization constants (K_b) for some amines are presented in Table 2.

Table 2 Ionization Constants of Various Amines[1]

Amine	K_b (mole l^{-1})
methylamine	4.4×10^{-4}
dimethylamine	5.2×10^{-4}
trimethylamine	5.5×10^{-5}
dodecylamine	4.3×10^{-4}
octadecylamine	4.0×10^{-4}

The ionization characteristics of the primary and secondary amines are very similar and changing the length of the alkyl group has little effect. In aqueous solutions the following equilibria occur:

$$RNH_{2\,(aq)} + H_2O \overset{K_b}{\rightleftharpoons} RNH_3^+ + OH^- \qquad (1)$$

$$RNH_{2\,(s)} \overset{K}{\rightleftharpoons} RNH_{2\,(aq)} \qquad (2)$$

The concentration of unionized amine at saturation decreases with increase in chain length and this is shown in Table 3 for the primary amines.

Table 3 Solubilities of Unionized Amines

Amine	K (mole l^{-1})
decylamine	5×10^{-4}
dodecylamine	2×10^{-5}
tetradecylamine	1×10^{-6}

Using equilibria (1) and (2) and mass balances concentration curves of the type shown in Figure 1 can be constructed. This shows that for dodecylamine at a total concentration of 10^{-4} mole l^{-1} pH substantially changes the form of dodecylamine in solution. RNH_3^+ predominates at pH values below 10 and at 10.6 the concentrations of RNH_3^+ and $RNH_{2\ (Aq.)}$ are equal. At higher pH values the unionized amine molecule precipitates as the concentration of $RNH_{2\ Aq}$ approaches its solubility (K). In this pH region the cationic amine concentration decreases markedly as more unionized amine precipitates. As will be shown later, the ionization characteristics of amines have a profound effect on the collecting properties of amines.

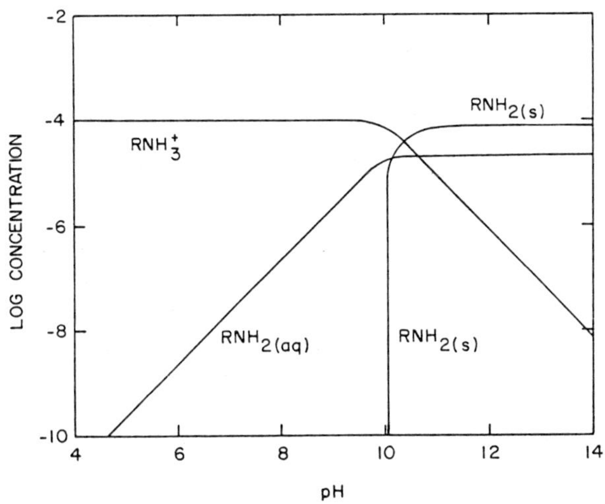

Fig 1 Logarithmic concentration diagram for 1 x 10^{-4} mole l^{-1} dodecylamine.

b) Micellization Micellization, a characteristic of heteropolar molecules such as alkyl chain based surfactants, assumes an important role in flotation systems. Micelles form because the hydrocarbon chain is non-ionic in nature, and a mutual incompatibility between polar water molecules and non-polar hydrocarbon chains exists. The forces which cause surfactant ions to associate at the cmc are partially responsible for the adsorption of surfactant at the mineral/water interface but at somewhat lower concentrations, as will be discussed below. As the alkyl chain length increases the cmc decreases (Table 4).

Table 4 Critical Micelle Concentration of Various Amines[3]

Amine	cmc (mole l^{-1})
decylamine	3.2×10^{-2}
dodecylamine	1.3×10^{-2}
tetradecylamine	4.1×10^{-3}
hexadecylamine	8.3×10^{-4}
octadecylamine	4.0×10^{-4}

There is a limit to the number of collector ions that can be contained within a micelle owing to the electrostatic repulsion between charged polar groups. Inorganic salts which produce ions of opposite charge, or neutral organic molecules, such as long-chain alcohols, reduce the repulsive forces between the charged heads and, hence, lower the cmc. In the case of dodecylamine chloride, the addition of 0.05 mole l^{-1} sodium chloride lowers the cmc from 1.3×10^{-2} to 6.8×10^{-3} mole l^{-1}.

The cmc of primary, secondary and tertiary amines will decrease as the pH is increased. The concentration of non-ionized amine will increase with increase in pH and co-micellize with the cationic amine ions. Eventually at high pH values when there is effectively no cationic amine present the micelles will be composed entirely of non-ionized amine molecules. For example, dodecylamine has a cmc of 1.3×10^{-2} mole l^{-1} at pH 6.0 and 10^{-4} mole l^{-1} at pH 11.0. [4].

Adsorption of Cationic Collectors at the Mineral/Water Interface

a) Mineral/Water Interface Most minerals when immersed in water acquire a surface charge because of the preferential adsorption of ions or the dissociation of surface groups. The presence of counter ions in the solution near to the solid leads to the establishment of the well known electrical double layer. The magnitude of the potential at the surface is dependent on the concentration of potential determining ions in solution. Generally, these ions are those which make up the crystal lattice but in many cases, as for oxides and silicates, H^+ and OH^- ions are potential determining. Thus for many minerals varying the pH modifies the sign and magnitude of the surface charge and at a certain pH, characteristic of the mineral, the net surface charge will be zero. The concentration of potential determining ion at which the net charge is zero is referred to as the zero point of charge (pzc) and typical values for some oxides and silicates are given in Table 5.

Table 5 The pzc of Some Oxides and Silicates

Mineral		pzc (pH)
quartz	SiO_2	1 - 2
cassiterite	SnO_2	4.5
rutile	TiO_2	5.8
hematite	Fe_2O_3	4.8 - 6.7
corundum	Al_2O_3	9.1
zircon	$ZrSiO_4$	5.8
beryl	$Be_3AL_2(Si_6O_{18})$	3.2
microcline	$K(AlSi_3O_8)$	1.8

At pH values below the pzc of oxides and silicates the minerals will be positively charged and above negatively charged.

For the salt type minerals such as calcite ($CaCO_3$), scheelite, ($CaWO_4$) and fluorite (CaF_2) the potential determining ions are those which make up the crystal lattice. However, the concentrations of lattice ion in solution will be dependent on the pH because the lattice ions are weak acids or bases so that H^+ and OH^- ions will also have a potential determining role. Thus, fluorite can be made positively charged by increasing the hydrogen ion concentration (pzc pH 6.2) or by increasing the calcium concentration (p Ca 2.6 - 7.7). Conversely increasing the pH and/or fluoride concentration will produce a negatively charged surface[6].

b) Adsorption of Cationic Surfactants Cationic surfactants are adsorbed at the mineral/water interface primarily by Coulombic attraction of the cationic surfactant ion for a negatively charged surface. In general therefore, adsorption takes place at pH values above the pzc of oxides and silicates and under analogous conditions for the salt-type minerals. The adsorption is reversible and gives a characteristic form of isotherm such as that shown for cetyl trimethyl ammonium bromide (CTAB) on rutile (TiO_2) at pH 10.5 in Figure 2[7].

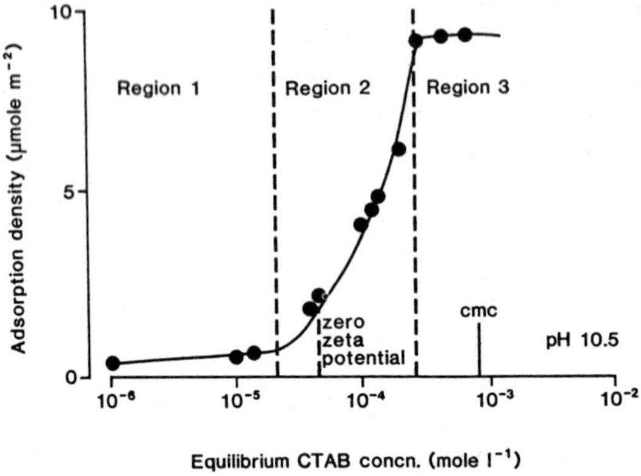

Fig 2 Adsorption of CTAB on TiO_2 at pH 10.5

From very dilute solutions the cationic surfactant ions are simply exchanged with cationic counter ions surrounding the negatively charged mineral particle (Region 1). The surfactant exhibits little specificity for the mineral surface over other monovalent cations. At somewhat higher concentrations - about two orders of magnitude less than the cmc - there is a remarkable feature in the isotherm, namely, sharp increase in affinity, which is due to lateral interactions of the adsorbed ions, or a form of condensation (Region 2). The effect is analogous to micellization in bulk solution and the two dimensional adsorbed associates have been termed hemimicelles[8]. Hemicellization leads to neutralization of the charge on the mineral surface and the formation of an adsorbed surfactant layer with a greater opposite charge than that on the mineral surface. The additional driving force for the adsorption is the formation of hydrophobic bonds between the non-polar parts of the surfactant ions. Adsorption continues until either a close-packed vertically oriented monolayer or bilayer is formed (Region 3). Figure 3 gives a schematic representation of the adsorbed ions in the three regions of the adsorption isotherm.

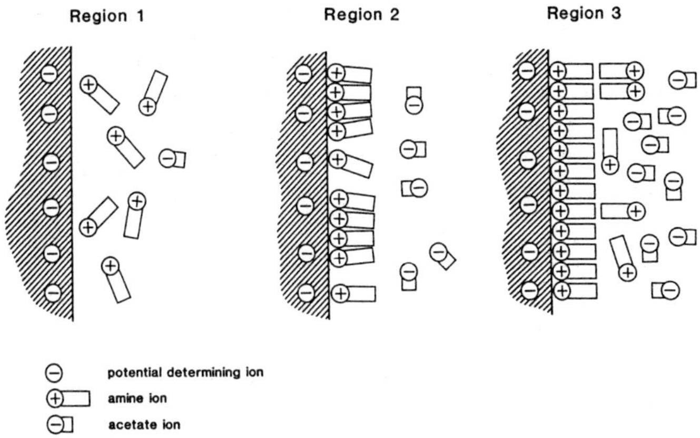

Fig 3 Schematic representation of amine adsorption

Flotation with Cationic Surfactants

Cationic collectors are used to render a mineral hydrophobic and hence floatable. Since this can only be achieved if the collector is adsorbed at the mineral/water interface it follows that flotation with cationic collectors will generally be obtained under conditions where the mineral is initially negatively charged. Furthermore factors which affect the concentration of cationic amine in solution such as ionization constant, solubility and cmc, which in turn are dependent on the size and nature of the hydrophobic group, will have a marked effect on the flotation response.

a) Effect of Collector Concentration and Hydrocarbon Chain Length

The onset of hemimicellization at the mineral/water interface gives rise to a marked increase in mineral hydrophobicity. This is demonstrated in Figure 4[9] where a comparison is made between the adsorption of dodecylamine at the quartz/water interface at pH 6 - 7 and the contact angle/flotation response obtained. The formation of hydrophobic associates begins at an equilibrium amine concentration of about 10^{-4} mole l^{-1}. To give the flotation response shown the adsorbed surfactant ions must be oriented as depicted in Figure 3. At higher amine concentrations approaching the cmc a bimolecular layer is formed with the hydrophilic groups oriented toward the aqueous phase. A decrease in hydrophobicity and hence flotation is therefore obtained and eventually at the cmc the quartz becomes hydrophilic.

Also shown in Figure 4 is the zeta potential of quartz as a function of amine concentration. Zero zeta potential is obtained at an amine concentration and adsorption density corresponding to an adsorbed layer with an equal and opposite charge on the mineral surface. At higher concentrations the quartz particle plus adsorbed layer acquires a net positive charge and at lower concentrations a net negative charge is obtained.

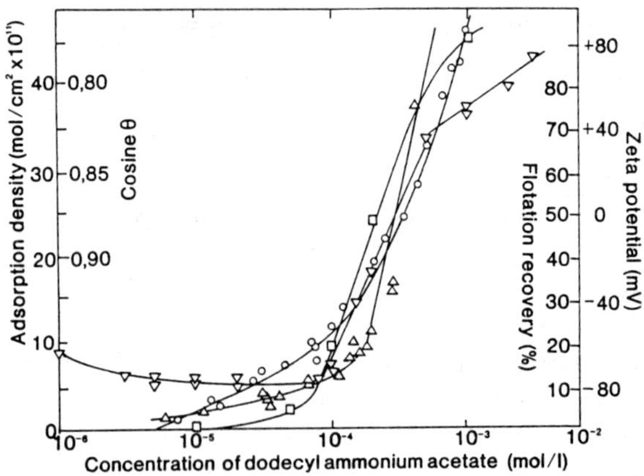

Fig 4 Adsorption density (Δ), contact angle (O), flotation response (□) and zeta potential (∇) for quartz as a function of dodecylamine concentration at pH 6 - 7.

Increasing the chain length of the cationic collector decreases the concentration at which hemimicelles are formed. It follows therefore that increasing the collector chain length should reduce the concentration of amine required to give flotation. Figure 5 [10] shows that this is the case. By defining the onset of hemimicelle formation by extrapolation of the sharp vertical rise in the flotation curve to zero flotation a plot of CHM (concentration at which hemimicelles form) against number of carbon atoms can be obtained. Such a curve is shown in Figure 6 together with a curve of cmc against chain length. The similarity in slope of these two curves is indicative that the forces responsible for micellization and hemicellization are similar. In fact it is possible to calculate from the slopes that the standard free energy for the removal of one mole of CH_2 groups from an aqueous environment is -1.1 kT [10]. Similar values have been obtained from other micellization studies.

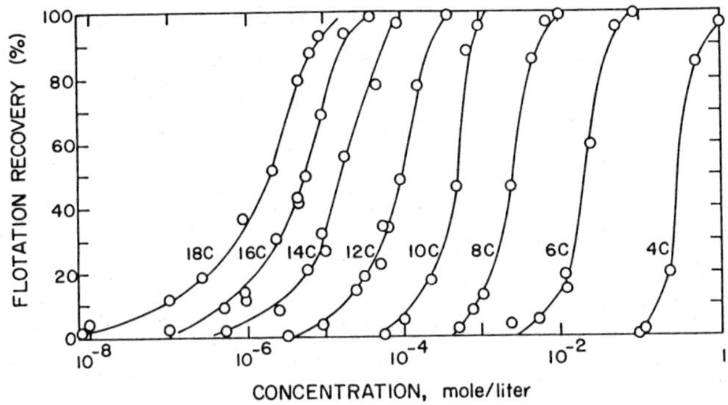

Fig 5 Flotation of quartz as a function of amine concentration and chain length at pH 6 - 7.

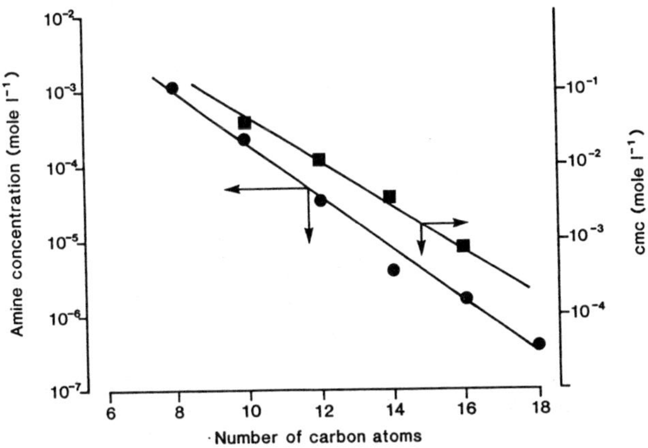

Fig 6 CHM of amines on quartz at pH 6 - 7 and the cmc as a function of amine chain length

b) Effect of pH An increase in pH above the pzc of oxides and silicates or in soluble lattice anion concentration for salt-type minerals will produce an increase in coulombic attraction for cationic amine ions. Increased adsorption and flotation response should therefore be expected. Typical results are shown in Figure 7[9] where flotation, contact angle and flotation data are shown for quartz in 4 x 10^{-5} mole l^{-1} dodecylamine solutions. The flotation recovery increases gradually as the pH is increased from the pzc. At pH values around neutral, however, there is a marked increase in the flotation recovery to a maximum between pH 9 and 10. This increase corresponds to the formation of hemimicelles at the quartz/water interface and hence to an additional mechanism of adsorption. Both the cmc and CHM will decrease as the pH is increased because of the incorporation of and coadsorption of non-ionized amine into the micelles and hemimicelles.

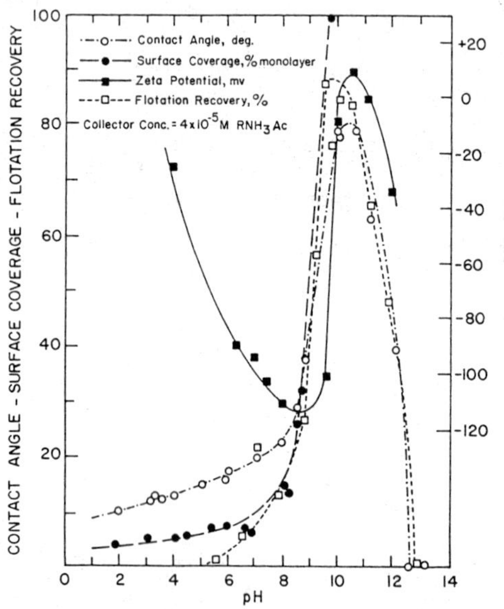

Fig 7 Contact angle, adsorption density, zeta potential and flotation of quartz as a function of pH in 4 x 10^{-5} mole l^{-1} dodecylamine.

Recovery decreases at high pH values because of the removal of cationic amine from solution as precipitated non-ionized amine. Maximum flotation recovery is usually obtained at the pH at which the primary, secondary and tertiary amines begin to precipitate. Under these conditions the non-ionized amine concentration will be at maximum (equal to the solubility K) and the cationic amine will predominate so that the coadsorption of non-ionized and cationic amine will be at a maximum. Non-ionized amine does not generally function as a collector.[11]

Flotation is possible with very long chain amines eg octadecylamine [12] at pH values just below the pzc of oxides and silicates (Fig 8). This is because although the surface charge is not positive there are still enough negative sites for the long chain amines to adsorb and produce a surface sufficiently hydrophobic for flotation. In general, such adsorption leads to a loss in flotation selectivity as the hydrophobic group in the collector is made larger.

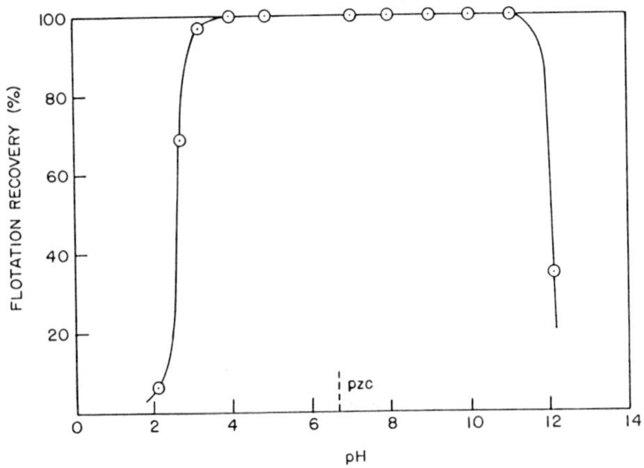

Fig 8 Flotation of hematite with 1 x 10^{-4} mole l^{-1} octadecylamine as a function of pH.

c) Effect of Amine Structure Although the structure of the non-polar part of amine collectors is extremely important very few quantitative data have been obtained under rigidly controlled conditions with sufficiently pure reagents. Studies of the dependency of cmc, solubility, ionization constant and flotation on structure are lacking. In going from a primary to tertiary amine by adding methyl groups to the nitrogen atom in say, dodecylamine a decrease in the solubility and cmc would be expected. The effect would not be large for a methyl group but would increase with the size of the substituted group.

Reference to the mechanism of amine adsorption at that the mineral/water interface suggests that maximum flotation will be obtained at slightly lower pH values as the degree of substitution is increased. That this is the case is demonstrated by the results summarized in Table 6[13].

Table 6 pH of Maximum Flotation, Precipitation of Nonionized Amine for Methyl Substituted Dodecylamine
(conc 4×10^{-4} mole l^{-1})

Amine	pH of ppt of Unionized Amine	pH of max. Flotation	Solubility of Unionized Species
Primary	9.4	9.0	2×10^{-5}
Secondary	9.5	8.8	1.2×10^{-5}
Tertiary	8.0	8.0	7.2×10^{-6}
Quarternary	-	11.3	-

Examples of Flotation With Cationic Collectors

Cationic surfactants, such as amines, are not very selective collectors because of their predominantly electrostatic mechanism of adsorption. Conditions under which just the required mineral in a multicomponent mixture is negatively charged are difficult, if not impossible, to achieve by adjustment of such solution parameters as pH. Instead selectivity is achieved by exploiting differences in magnitude of mineral surface charge and by the use of additional reagents known as depressants and activators which enhance the differences in surface properties of the minerals. Depressants are reagents added to prevent the flotation of unwanted minerals and activators are used to promote the adsorption of collector on a given mineral. For example in the flotation of siliceous minerals from iron oxide minerals, starch is commonly added to prevent the flotation of the iron oxides; and in the flotation of beryl and feldspar from quartz at pH 3 sodium fluoride is used to activate the beryl and feldspar. Detailed discussion of the mechanisms of activation and depression can be obtained elsewhere[14].

Amines used commercially are generally primary aliphatic types with C12 to C18 chain lengths although secondary amines with a shorter, say propane, chain with the longer chain are also used. Tallow amine derived from tallow fatty acid and composed of 18C unsaturated and saturated and 16C saturated amines is a common collector. Another common collector is cocoamine obtained from coconut oil fatty acids. Here 12C primary amine predominates with small amounts of 8C and 18C amine. Amine collectors are generally sold as the unreacted amine or as the acetate or chloride.

a) <u>Treatment of Iron Ores</u> The most important sources of iron are the minerals hematite (Fe_2O_3), magnetite Fe_2O_3, goethite FeO.OH, and siderite $FeCO_3$. Although there are ores in which one only of these minerals occurs, it is common-place to find the minerals in an intimate mixture. Associated gangue minerals are primarily quartz and silicates. When the iron minerals exhibit a high magnetic susceptibility the ore can be treated by magnetic separation. There are cases, however, where the iron minerals are not magnetic and this is especially true of the oxidized taconites in the USA.

The oxidized taconites in the Lake Superior district consist of hematite and goethite in intimate association with mainly quartz (jasper). Liberation of the iron oxides is obtained by grinding to at least 85 per cent finer than 25 μm which generates some extremely fine material (slimes) which cannot be treated by direct flotation. The process developed by the US Bureau of Mines and applied by the Cleveland-Cliffs Iron Co. at their Tilden mine consists of fine grinding, a fine degree of mineral dispersion, selective flocculation of the iron oxides, desliming to remove the dispersed silica-bearing slimes followed by the flotation of the silicates, not removed in the desliming stage, from the flocculated iron oxides. The process is represented schematically in Figure 9[15].

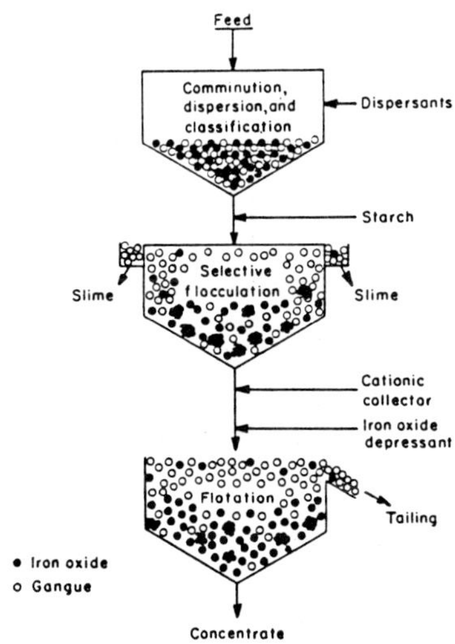

Fig 9 Schematic representation of the selective flocculation/flotation process for the treatment of iron ores.

Caustic soda and sodium silicate are added at the grinding stage to ensure that the pulp is dispersed. Selective flocculation of the iron oxide minerals is achieved by the addition of starch. Non-ionic corn, pearl or tapioca flour starch is used. The cationic and anionic varieties produce non-selective flocculation. Coarse silica is floated from the flocculated iron oxides with amine based collectors. Primary alkyl amines and alkyl/oxypropyl amines are used. Additional starch is added at the flotation stage to improve the depresion of the iron oxide minerals.

The Tilden operation started in 1974 and now produces 8.2 million tons per annum of pellets at 64 per cent iron. Sodium hydroxide and sodium silicate consumptions are of the order of 1 kg $tonne^{-1}$ with about 0.2 kg $tonne^{-1}$ of starch and amine collector.

b) Flotation of a Pegmatite The mineral beryl ($Be_3\ Al_2\ [Si_6O_{18}]$) occurs in granite pegmatites in association with mica, quartz and feldspar. Minor minerals also present may include spodumene, garnets, tourmaline, topaz and iron oxides. A possible flowsheet to treat such a material is shown in Figure 10. An understanding of the conditions used in each stage can be obtained by assuming the following pzc values for the major components present; beryl pH 3.5, quartz pH 2.5, feldspar pH 2.5, mica pH 1 and iron oxides pH 6.

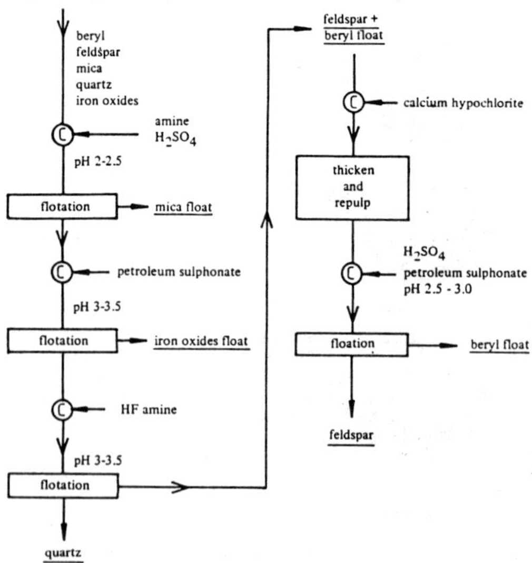

Fig 10 Treatment of a pegmatite by flotation

In the first stage, mica is floated with amine collectors at pH values between pH 2 and 2.5. Under these conditions, the mica is negatively charged, but all the other minerals have zero or positive charges. Next, the iron oxides are floated with anionic collectors such as C12 to C18 alkyl sulphonates or their equivalents at pH 3 to 3.5. The third stage consists of conditioning the pulp with HF or sodium fluorosilicate and adding more amine collector, at pH 3 to 3.5. The HF activates both the beryl and feldspar ie increases its negative charge so that both minerals float, leaving quartz as the tailing.

Separation of the beryl and feldspar can only be achieved after the amine present on both minerals is removed. This can be achieved by oxidative washing with calcium hypochlorite solution and then adding a petroleum sulphonate collector at pH 2.5 - 3.0 whereupon the beryl is floated away from the feldspar.

There are many possible variations to the flowsheet depending on the major and minor components of the pegmatite. Flotation with amines is used in a number of processes to recover feldspar, mica, quartz, and spodumene from non-beryl containing pegmatites. Amine consumptions in these processes are generally below 0.1 kg/tonne of ore treated.

c) <u>Flotation of Soluble Salts</u>

The concentration of sylvite (potash) from halite is a unique flotation process because it takes place in a saturated solution of the two soluble minerals. The ore is crushed and ground and suspended in a saturated brine solution. Prior to flotation the pulp is deslimed to remove fine clay minerals which would consume large quantities of collector and reduce the sylvite concentrate grade. Starch or guar gum is often added to prevent residual silicates from adsorbing collector.

Cationic collectors such as tallow amine acetates are used to float the sylvite (KCl) away from halite (NaCl). Consumptions range from 0.07 to about 0.2 kg/tonne of ore. The flotation is very selective and concentrate grades of 95 to 99 per cent are often obtained at high recoveries. The amine collector must, in general, contain at least 8 carbon atoms in the chain and similar to the flotation of oxides, silicates etc. the pH must correspond to the presence of cationic amine. Flotation is not obtained with quaternary compounds.

This process is very remarkable because it involves the separation of two isomorphous minerals which differ only in the metallic cation. It is even more remarkable when it is appreciated that a cationic collector is used to distinguish between two alkali metal salts. One hypothesis suggests that the reason for the selectivity between sylvite and halite is that the cationic polar group of a primary amine has similar dimensions to that of a K^+ lattice ion so that it can occupy this position at surface sites. A similar occupation on the surface of halite is not sterically favourable. This hypothesis is probably too simplistic because it does not explain flotation with other types of collector. [16]

Cationic collectors have a wide application in the flotation of non-sulphide minerals. Other examples include the flotation of quartz and silicates from calcium carbonates (calcite, limestone and chalk) and phosphate ores, and the flotation of the oxidized zinc minerals smithsonite, hemimorphite and wiltemite.

Summary and Conclusion

The main use of cationic surfactants in mineral processing is as a collector in the flotation process where they are used to float non-sulphide minerals. Primary amines are the principal reagent and their application is dependent on both the solution properties of the collector and the surface chemistry of the minerals. Best flotation is obtained where the collector is in the cationic form and the mineral is negatively charged. The chain length of the collector is generally equivalent to an alkyl chain containing between 10 and 20 carbon atoms.

Acknowledgement

The author wishes to thank the British Petroleum Company plc for permission to publish this paper.

References

(1) A.M. Gaudin "Flotation", McGraw-Hill, New York. 2nd Ed. 1957 p. 219.

(2) D.W. Fuerstenau (Ed.) "Froth Flotation" 50th Anniversary Volume, AIME, New York, 1962, p. 170.

(3) A.W. Ralston. "Fatty Acids and Their Derivatives", John Wiley and Sons, New York, 1948.

(4) D. Watson and R.M. Manser 'Trans. I.M.M. (Sect C: Mineral Process Extr. Metall). 1968, 77 c57-60.

(5) D.W. Fuerstenau Pure Appl. Chem. 1970, 24, 135-164.

(6) H.S. Hanna and P. Somasundaran 'Flotation' A.M. Gaudin Memorial Volume, edit. M.C. Fuerstenau, AIME, New York 1976, 1, 197 - 272.

(7) R. Stratton-Crawley and H.L. Shergold Colloids and Surfaces 1981, 2, 145-154.

(8) A.M. Gaudin and D.W. Fuerstenau Trans. AIME, 1955, 202, 958 - 962.

(9) D.W. Fuerstenau Min. Engng. 1957, 9, 1365-1367.

(10) D.W. Fuerstenau, T.W. Healy and P. Somasundaran. Trans. AIME 1964, 229, 321.

(11) H.L. Shergold and O. Mellgren Trans AIME, 1970, 247, 149-159.

(12) I. Iwasak, S.R.B. Cooke and H.S. Choi Trans AIME 1960, 217, 237.

(13) R.W. Smith and S. Akhtar. 'Flotation' A.M. Gaudin Memorial Volume. Ed. M.C. Fuerstenau. AIME, New York 1976, 1, 87-116.

(14) H.L. Shergold 'The Scientific Basis of Flotation' Ed. K.J. Ives, Martinus Nijhoff Publ. The Hague, 1984 p. 229-287.

(15) A.F. Colombo. 'Fine Particle Processing'. Ed. P. Somasundaran. AIME, New York 1980. 2, 1034-1056.

(16) J. Leja. 'Surface Chemistry of Froth Flotation', Plenum Press, New York, 1982, 536-541.

Organophilic Clays

By J. P. Tatum
PERCHEM LIMITED, WEST ROAD, TEMPLE FIELDS, HARLOW, ESSEX CM20 2BU, UK

Introduction

An organophilic clay is the product of an ion exchange reaction between a clay having an exchangeable cation on its surface and an organic cation. The properties of the clay can be radically altered by such an exchange. A swelling clay will change; from being hydrophilic, swelling in water to give a gel structure, the clay will become hydrophobic and will be capable of swelling in organic liquids to give a gel.

The organic cations used commercially in the ion exchange process are almost exclusively quaternary ammonium compounds with one or two long chain alkyl or aryl groups, i.e. they are cationic surfactants.

An indication of the importance of the organophilic clay industry as a market for quaternary ammonium compounds can be obtained from production estimates which range from 30,000 to 40,000 tonnes per annum worldwide[1]. The organic cation content of an organophilic clay will usually be between 35 and 45% w/w, thus, using the estimated production figures above, the worldwide annual usage of quaternary ammonium compounds (on a 100% basis) is between 10,500 and 18,000 tonnes.

The Clays

In order to produce an organoclay which will have useful gel forming properties, it must be swellable and have a sufficiently high cation exchange capacity to enable a useful quantity of the organic cation to react. In practice, the clays used are bentonites.

The nomenclature of clays is frequently confusing. The following definition of bentonite by Ross and Shannon[2] while not all embracing is useful: "Bentonite is a rock composed essentially of a crystalline clay like mineral formed by the devitrification and the accompanying chemical alteration of a glassy igneous material, usually a tuff or volcanic ash". The term bentonite is now used to denote any clay which is chiefly composed of a smectite clay mineral and which relies on the smectite clay for its properties[3].

Smectite clays are three sheet, or trimorphic, minerals composed of an octahedral di or trivalent metal oxide layer sandwiched between two sheets of linked silica tetrahedra (figure 1).

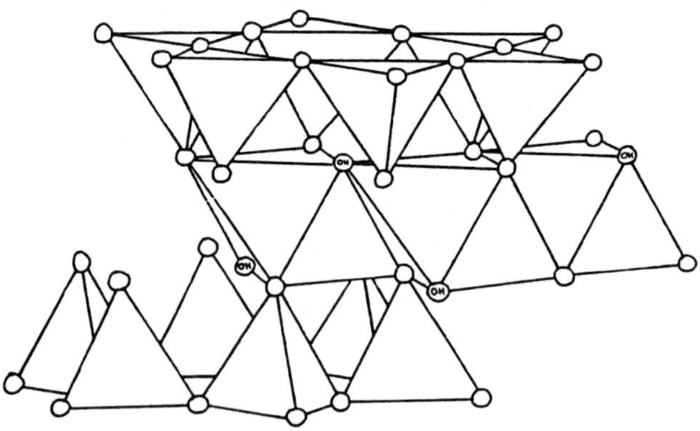

FIGURE 1: The trimorphic layer structure of smectite clays.

The valency of the metal ion in the octahedral layer gives rise to two sub-groups[4]. If the metal is divalent all three octahedral positions in a unit cell are filled and the mineral is referred to as trioctahedral (figure 2a). If the metal is trivalent only two out of three octahedral positions are filled and the mineral is dioctahedral (figure 2b).

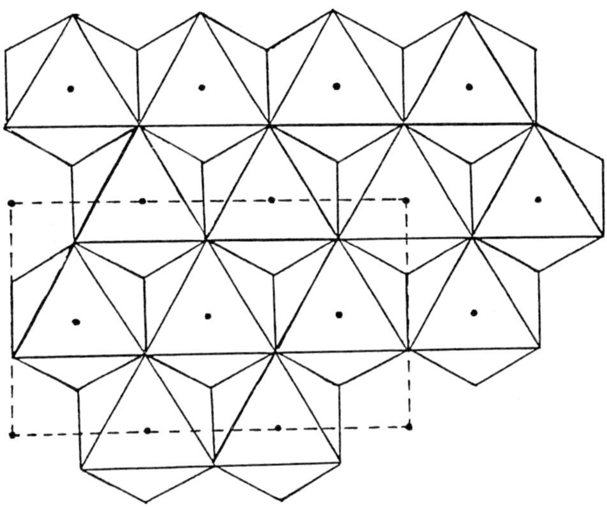

FIGURE 2a: Trioctahedral layer structure.

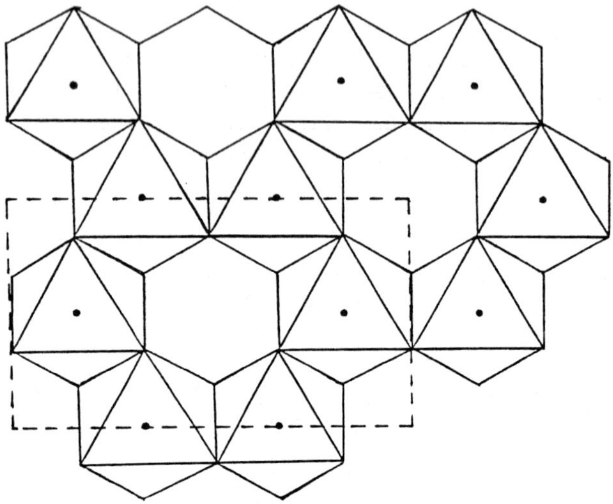

FIGURE 2b: Dioctahedral layer structure.

Isomorphous replacement in both the octahedral and tetrahedral layers by cations of similar size but different valency is possible. This type of substitution frequently gives rise to a charged structure[5]. Internal rearrangements will neutralise some of this charge, but in the majority of cases the charge is balanced by the adsorption of cations between the layers. Substitution gives rise to series of minerals, for example, the montmorillonite-beidellite series:[6]

$$[(Al_xMg_y)_2(Si_{4-z})O_{10}(OH)_2]M^+_{y+z}$$

Where in the case of montmorillonite z=0 and in the case of beidellite y=0.

In addition to charges caused by isomorphous cation substitution, a smaller proportion of charges are caused by broken bonds at the edges of the clay platelet[7], again some of the edge charges are neutralised by rearrangements and some by cation adsorption. The contribution of edge charges to the cation exchange capacity will vary with pH.

Organic Complexes

Organic complexes can be made with clay minerals in a number of ways. All interlayer complex formation relies on diffusion of the organic species through the clay layer structure. After which it may form a complex either, in the case of a noncharged, polar molecule, by competition with water for ligand positions around the exchangeable cation[8], or, in the case of organic cations, by an ion exchange process.

Cation exchange is almost instantaneous where an exchange reaction occurs with kaolinite where the exchange capacity is limited to the edge of the particles[9] and no interlayer penetration occurs. In montmorillonite, however, where the majority of the exchangeable cations are held between the layers the rate is diffusion controlled.

Measurements of diffusion rate on montmorillonite are difficult, due to the small grain size, but measurement in vermiculite is possible using optical microscopy to follow increases in basal spacing in single flakes[10]. The replacement of magnesium by strontium is accompanied by an increase in spacing from 1.44 to 1.5 nm. The rate was found to increase with temperature and to decrease with increasing charge density. Montmorillonite has a lower charge than vermiculite and so diffusion is faster. Jordan[11] reacted Wyoming bentonite with n-alkylammonium cations and measured the basal spacings of the resulting organoclays using x-ray diffractometry. He found that the basal spacing remained constant at approximately 1.36 nm when the alkyl chain length was between three and ten carbon atoms. The interlayer spacing is approximately 0.4 nm which is equivalent to the van de Waals thickness of a methyl or methylene group in a planar orientation to the clay surface. Beyond C_{10}, up to a chain length of C_{18} the basal spacing was increased to approximately 1.76 nm. Assuming a planar orientation, Jordan calculated the area covered by the amines used and expressed them as a percentage of the area covered (Table 1).

Carbon chain length	001 spacing nm	Clay platelet separation nm	No. of amine layers	Calculated amine area nm^2	Clay coverage %
0	0.96	-	-	-	-
3	1.35	0.39	1	0.38	23
4	1.34	0.38	1	0.44	27
8	1.33	0.37	1	0.69	42
10	1.36	0.40	1	0.81	49
12	1.74	0.78	2	0.94	57
14	1.74	0.78	2	1.06	64
16	1.75	0.79	2	1.19	72
18	1.76	0.80	2	1.31	79

TABLE 1: Effect of increasing alkyl chain length on inter-layer spacing.

As the area covered by the ammonium cation exceeds 50%, it is no long possible for the organic cations on opposing surfaces of the clay to fit in a unimolecular layer between the platelets and so the basal spacing increases by an amount equal to that of another alkyl layer. Jordan found that this was also the point at which the organoclay started to swell appreciably in organic liquids (Graph 1).

In later work Jordan et al.[12] investigated the basal spacing of montmorillonite complexes with increasing amounts of octadecyl-amine (Graph 2).

GRAPH 1: Effect of alkyl chain length on swelling of organoclays in nitrobenzene.

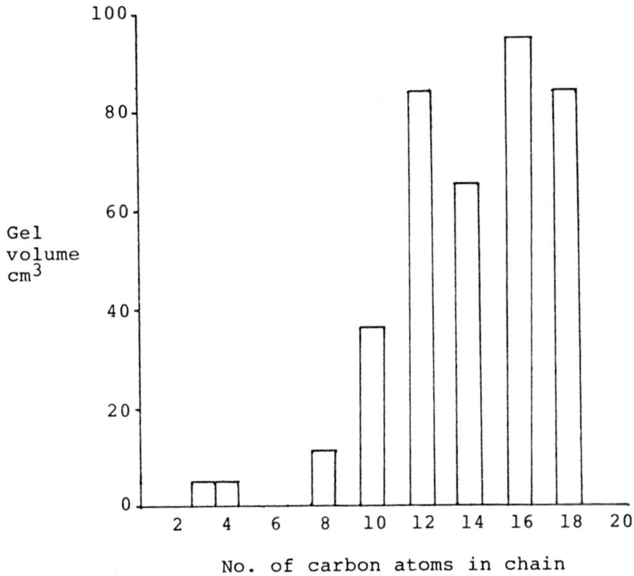

GRAPH 2: Effect of amine content on platelet separation.

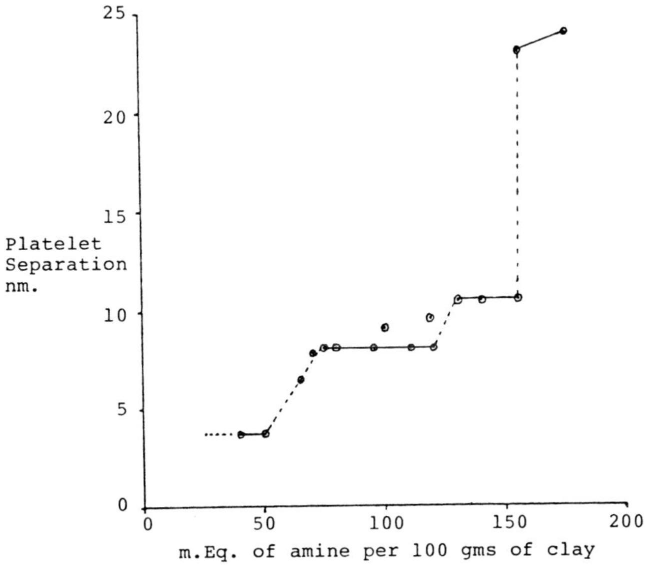

The lower plateau, corresponding to a separation of 0.4 nm, is assumed to be the situation where less than 50% of the surface is covered, a stepwise increase up to a separation of 0.8 nm occurs when this coverage is exceeded. A further increase occurs when no further amine can be accommodated on the surface of the platelets in a planar conformation, and the amine chains orientate themselves on edge. The effective width of a zigzag alkyl chain is 0.49 nm giving a theoretical separation of 0.98 nm, in fact, a spacing of 1.04 nm was observed. Beyond this plateau continuous d spacings were observed up to 2.4 nm which were attributed to the amine chain standing at right angles to the platelet surface. The effect of the amine/clay ratio on swelling was also investigated (Graph 3).

GRAPH 3: Effect of amine content on clay swelling in nitrobenzene.

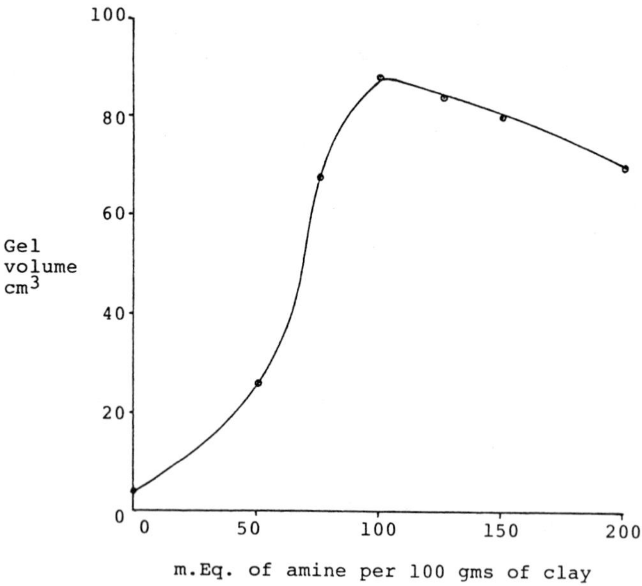

A wide variety of quaternary ammonium compounds have been used to make organoclays, but, chiefly for economic reasons, only a few are used in commercial processes. Of these few, dimethyldioctadecylammonium (figure 3a) and dimethyloctadecylbenzylammonium (figure 3b) cations are the most commonly used.

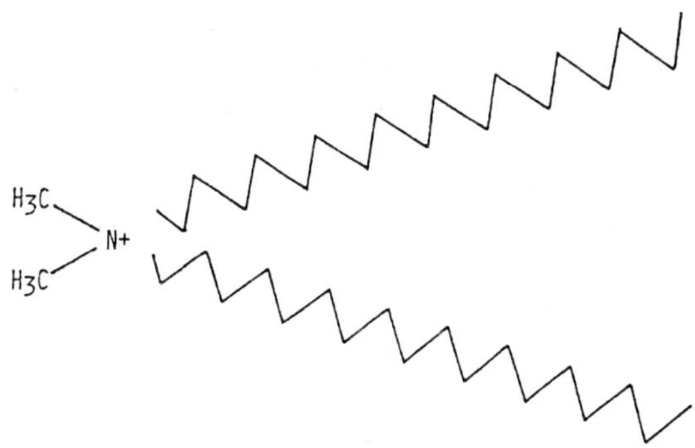

FIGURE 3a: Dimethyldioctadecylammonium cation.

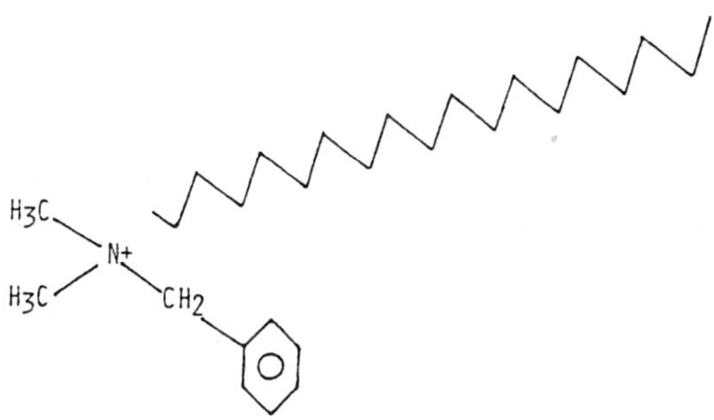

FIGURE 3b: Dimethyloctadecylbenzylammonium cation.

The octadecyl chain is derived from hydrogenated tallow and for brevity, the materials will be referred to as DMDHT and DMHTB ammonium cations.

The manufacture of an organophilic clay is essentially a simple process. The clay, which may undergo pretreatment, is slurried in water at a concentration of not more than ca. 10% w/w. After allowing adequate time and agitation for substantially complete hydration and separation of the platelets the slurry is purified by centrifuging. This removes the non-clay materials present as well as some non-swelling clay. The clean clay slurry is then agitated while the liquid quaternary ammonium salt is slowly added. Good agitation is necessary as the viscosity of the reaction mix increases dramatically up to the equivalence point due to agglomeration of the platelets caused by the reduction in surface charge. After the equivalence point, further addition causes the platelets to become covered with the positively charged organic species and the platelets will once more repel each other, this leads to deagglomeration and a reduction in viscosity. Although the reaction is very fast, the cations are still mobile on the clay surface, and additional time in the reaction vessel gives an improvement in product quality.

The reaction mix is easily filtered as the clay is now strongly hydrophobic, the filter cake is then washed and the organoclay dried and milled.

Gel Formation

An organophilic clay can be easily dispersed in most organic liquids using high shear mixing; however, gellation will only occur if the organic liquid system contains a mixture of polarities. In some cases one molecule may suffice: for example, in his original work on the swelling of organoclays Jordan used nitrobenzene as the organic liquid[12]. In other cases it may be necessary to add a small quantity of a highly polar molecule to the organoclay/organic liquid system, for example, ethanol, propanone, water etc. The effect of such a polar additive has been attributed to the polar molecule solvating the clay surface which is not covered by the organic cation[13]. The alkyl group of the cation then interacts by means of van der Waals forces with the non-polar component of the organic liquid. The concentration of the polar additive has a very profound effect on gel strength which increases with increasing polar additive concentration up to a maximum, after which further additions give a reduction in gel strength (Graph 4).

GRAPH 4: The effect of polar additive concentration on viscosity (5% DMDHT organoclay in mineral oil).

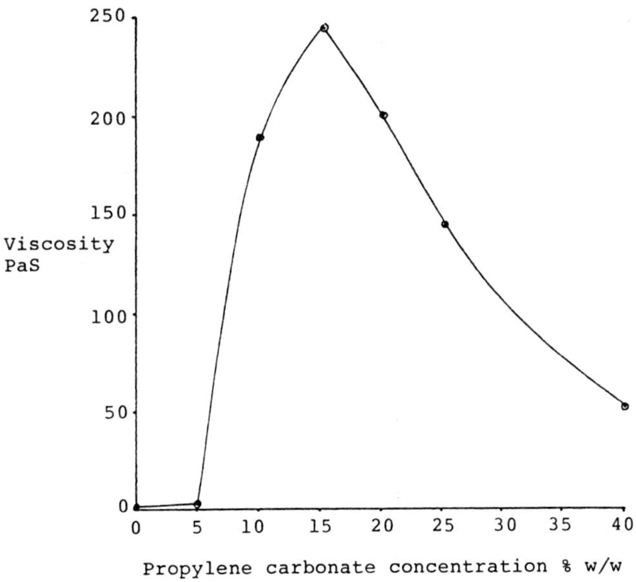

This phenomenon was investigated by Granquist and McTee[14]. They suggested a theory whereby the polar molecule adopts a preferred orientation across the organoclay/organic liquid interface and promotes dissociation of the quaternary ammonium montmorillonite complex. This action results in the development of a diffuse double layer around the clay platelet. Thus as the concentration of the polar additive increases the level of gellation increases up to a maximum beyond which the polar additive promotes redispersion of the clay and an accompanying reduction in viscosity.

Applications Of Organoclays

Greases

The simplest industrial application of organophilic clays is the production of a lubricating grease. In this application a lubricating oil is gelled with between 5 and 12% w/w of an organoclay usually based on DMDHT which is more suitable for the mainly aliphatic oils used for lubrication. Where an oil is used with a high naphthenic content a DMHTB based organoclay may be used.

One facet of the performance of organoclay gelled greases which distinguishes them from other types, e.g. soap based greases, is their non-melting character. A knob of grease can be placed on a palette knife and will remain in its original position, despite being ignited and allowed to burn. This behaviour is obviously very useful in high temperature applications and can permit the use of a thinner grease than would be the case with more heat sensitive gellants.

Paints

Modern oil based paints have been developed from old varnish technology. In many respects there is very little difference between a paint where the pigment is litharge, the binder is linseed oil and the solvent is turpentine and a modern decorative paint where the pigment is a coated titanium dioxide, the binder is synthetic alkyd resin and the solvent is white spirit. The principal differences between the two paints will be due to the methods by which they are made and the additives used. The old paint would probably have had the pigment ground in using a mill, whereas the pigment used in the modern paint is easily dispersed using a high speed stirrer. The uniformity introduced by the use of synthetic resins in conjunction with well characterised metal soap driers means that the modern paint will dry faster and last longer than the old paint.

The use of organophilic clays in very small quantities (between 0.25 and 1.0%) will reduce pigment settling, liquid separation and enhance the "customer appeal" of the product in the can. Application benefits are also provided as sag and "curtaining" are reduced, and absorption into porous surfaces is controlled.

The decorative paints described are closely linked to the varnishes from which they derive, but paints for industrial use may be based on a wide variety of solvent and resin types. The white spirit based paints are most effectively viscosified using DMDHT organoclays. As the polarity of the solvent base increases through to aromatic solvents which are used in, for example, chlorinated rubber paints and then to paints based on oxygenated solvents, for example, automotive cellulose lacquers, the preferred gellant becomes a clay modified with DMHTB. Performance criteria will dictate which gellant to use as other components in the paint system can have a profound effect on the properties producted by the organophilic clay.

Inks

Organophilic clays are used in rotary letterpress printing inks to reduce a problem known as ink fly or misting. This occurs when ink is pinched between the ink roller and the transfer roller (Figure 4) causing the ink to string. The string produces very small droplets of ink when it breaks up and these droplets pose a respiratory hazard. The organophilic clay reduces this by shortening the length of the string so that when it breaks fewer droplets are produced. The addition of organophilic clay permits high printing speeds and gives cleaner print room conditions.

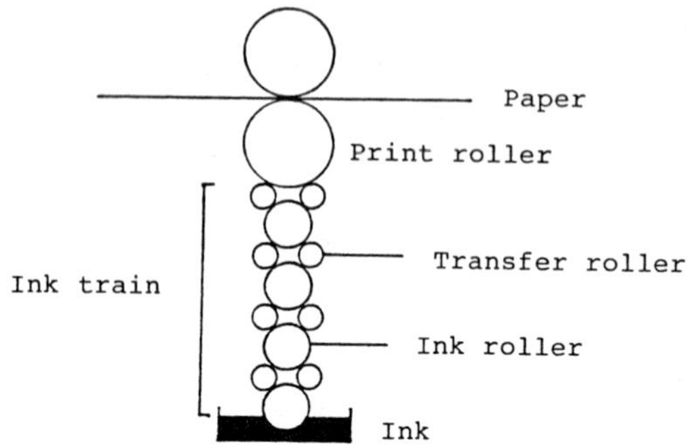

FIGURE 4: Diagram of rotary letter press printing equipment.

Oil Well Drilling Muds

The drilling mud used in drilling oilwells has five main functions; it must

1. Cool and lubricate the drill bit and string.
2. Help suspend the drill string.
3. Suspend drill cuttings - especially when drilling is interupted.
4. Stabilise the formation.
5. Minimise losses into the formation.

The relative importance of these functions will depend on the formation, the depth of the well and the angle of drilling (figure 5).

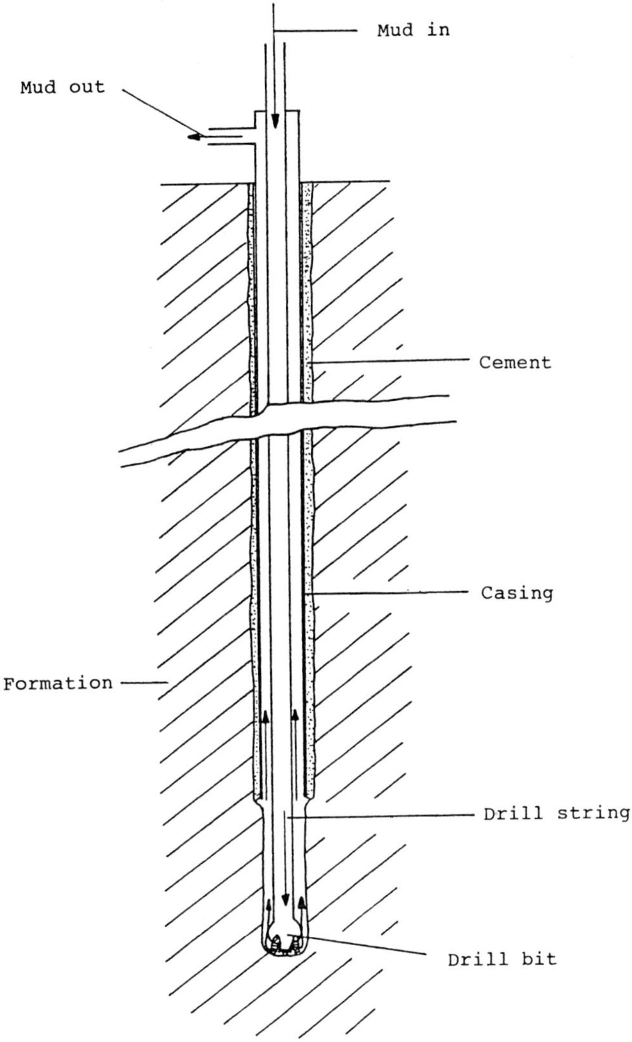

FIGURE 5: Drilling mud flow during drilling.

Most drilling muds are based on water which is viscosified either with bentonite clay or a water dispersible polymer. In some wells especially very deep, hot or highly deviated wells, or when drilling through water sensitive formations, the use of a mud based on a water in oil emulsion will be advantageous. Originally oil based muds were composed of water or brine emulsified in diesel oil. A primary emulsifier was used to produce the emulsion, which is viscosified and made thixotropic by the addition of a DMDHT based organophilic clay. Barytes is added as a weighting agent to increase the density of the mud which will help suspend the drill string. To ensure that the barytes and drill cuttings are oil wet a secondary emulsifier is added.

The organophilic clay has many functions in the mud. The thixotropy imparted to the mud will suspend the cuttings if the mud flow is interrupted. It will stabilise the water in oil emulsion, and it reduces mud losses into the formation by forming a thin impermeable filter cake on the walls of the hole.

In recent years growing pressure to reduce the impact of drilling fluids on the environment has led to changes in the formulation of oil muds, particuarly those used for offshore drilling. The main change is the substitution of oils containing low levels of aromatics for diesel to give low toxicitiy oil muds. The changes in oil have led to changes in the organophilic clays used. Diesel is far more effective at solvating the organic chains of the organophilic clays than are the new non-toxic oils. The result of this is that conventional DMDHT montmorillonites only develop their full performance after very high shear mixing, often only after they have been circulated through the drill bit a few times. New organophilic clays based on quaternary ammonium compounds with a greater organic content, for example methylbenzyldioctadecylammonium or methyltrioctadecylammonium clays, will give good rheological properties immediately the mud is made-up.

Discussion

The applications mentioned are only four of many such uses. It is possible to introduce viscosity into many different types of solvent systems using organophilic clays.

As mentioned earlier, the polarity of the solvent is very important in the choice of which organic cation to use on the clay. A large proportion of aliphatic content in the organoclay will favour its uses in low polarity solvents, a reduction in the alkyl chain length or numbers will favour its use in increasingly high polarity solvents. The substitution of a benzyl group for a hydrogenated tallow chain will also favour the use of the organoclay in high polarity solvents.

Although many different cations can be used and new cations are continually being evaluated, in practice the number employed is quite small as many improvements which may be made by substitution of organic cations are negated by the increased cost of the finished product. In addition, the performance of the organoclay is dependent on the formulation into which it is put and changes in the formulation and in the processing of the organoclay can enable organoclays to be used in many seemingly unfavourable systems. For such systems experience and experiment are necessary to obtain the ultimate performance from the organoclay.

Conclusion

Organoclays are a unique type of viscosifier, performing in many cases, a unique function. As the chemistry of clays and clay-organic interactions becomes more fully understood, improvements in performance and more diverse applications can be expected to take place.

References

1. G.M. Clark (Ed) Ind. Minerals, 1985, 216, 39.
2. C.S. Ross and E.V. Shannon, J.Am. Ceram. Sci., 1926, 9, 77-96.
3. R.E. Grim, N. Güven, Bentonites: geology, mineralogy, properties and uses, Elsevier Scientific Publishing Company, Amsterdam (1978) p.1.
4. E. Nemecz, Clay Minerals, Akademiai Kiado, Budapest (1981) pp.25-26.
5. B.K.G. Theng, The Chemistry of Clay-Organic Reactions, Adam Hilger, London (1974) p.2.
6. R.E. Grim, N. Güven, Bentonites; geology, mineralogy, properties and uses. Elsevier Scientific Publishing Company, Amsterdam (1978) p.5.
7. R.E. Grim, Clay Mineralogy, McCraw-Hill, New York (1953) p.194
8. B.K.G. Theng, The Chemistry of Clay-Organic Reactions, Adam Hilger, London (1974) p.20.
9. R.E. Grim, Clay Mineralogy, McCraw-Hill, New York (1953) p.200.
10. G.F. Walker, Nature, 1959, 184, 1392-1393.
11. J.W. Jordan, J.Phys. Colloid Chemistry, 1949, 53, 294-306.
12. J.W. Jordan, B.J. Hook and C.M. Finlayson, J.Phys. Colloid Chemistry, 1950, 54, 1196-1208.
13. B.K.G. Theng, The Chemistry of Clay-Organic Reactions, Adam Hilger, London (1974) p.232.
14. W.T. Granquist and J.L. McTee Jnr. J. Colloid Sci. 1963, 18, 409-420.

Antistatic/Conductivity Additives in Paint and Plastics

By B. Davis
ABM CHEMICALS LIMITED, WORTLEY LOW MILLS, WHITEHALL ROAD, LEEDS LS12 4RF, UK

Conductivity Additives for Spray Paints

In the electrostatic spray technique, the atomised paint particles receive a high voltage in the "gun". Subsequently the paint particles are attracted to the earthed work-piece, resulting in highly efficient paint utilisation with good coverage of any awkward areas (Figure 1).

fig.1

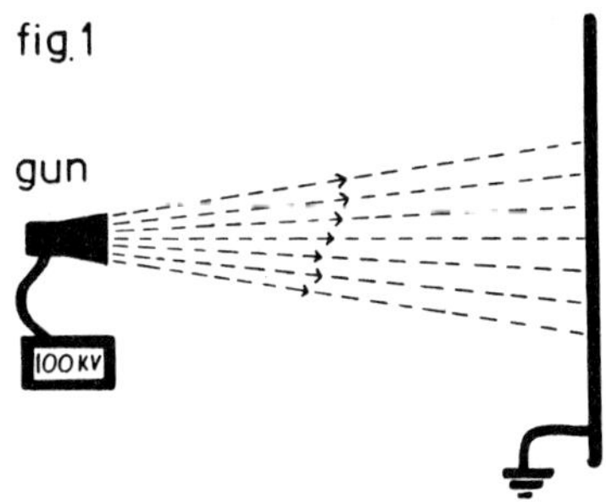

Paints formulated largely from non-polar solvents have an electrical resistance which is far too high for satisfactory application in this way. Reformulating the paint using a high proportion of polar solvent is one method of overcoming this difficulty, as illustrated by the effect on resistance as butanol is added to xylene (Table 1).

Table 1. The resistance of xylene/butanol mixtures

% Butanol	Resistance MΩ
0	> 10,000
10	2,000
20	100
30	8
50	0.5

It is necessary to reduce the paint resistance below 1 megohm so very large proportions of polar solvent would be necessary, therefore a search for a more efficient chemical additive is illustrated in the following selected list. (Table 2).

Table 2. The resistance of xylene containing 4% additive.

Additive	Resistance MΩ
None	>10,000
Ethoxylated fatty amine (2EO)	6,000
Ethoxylated fatty amine (15EO)	200
Benzyl blocked non-ionic	100
Butylamine dodecyl benzene sulphonate	60
Fatty imidazoline	8
Fatty imidazoline salt	4
Anionic phosphate ester	0.7
Fatty quaternary ammonium salt	0.5

Some of the more hydrophilic molecules were not completely soluble and apart from an anionic phosphate ester which might be useful in special cases, where compatiblity problems exist, the cationic quaternary compounds are of greater interest and a more detailed examination was carried out to try and obtain the optimum chemical structure (Figure 2).

Quaternary ammonium compounds containing small alkyl groups were not particularly effective but this appeared to be mainly because of the low solubility in the xylene solvent. Varying the nature of the anion did affect the result as it changed the solubility (Table 3).

Figure 2. Structure of quaternary ammonium compounds

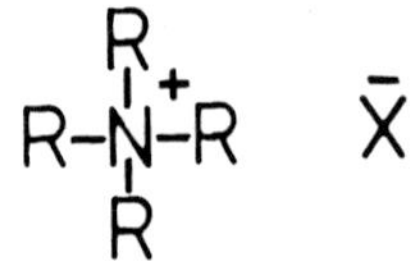

Table 3. The resistance of xylene containing 4% additive

Alkyl Group R	Anion $\bar{X}$	Resistance MΩ
Ethyl	$C_2H_5SO_4$	> 10
Hydroxy ethyl	$C_2H_5SO_4$	> 10
Butyl	Br	> 10
Butyl	BF_4	8
Butyl	$C_2H_5SO_4$	0.6

Better results are obtained when at least one of the alkyl groups is large. The nature of the anion appears to be important only if it results in a compound of low xylene solubility, which can occur with quaternary bromides or chlorides. Halogen ions are in any event undesirable because of their tendency to encourage corrosion under the paint film (Table 4).

Table 4. The resistance of xylene containing 4% higher alkyl additive

	Resistance MΩ	Solubility
$C_{14}H_{29}-\overset{+}{N}(CH_3)_3 \cdot Br^-$	> 10	Very cloudy
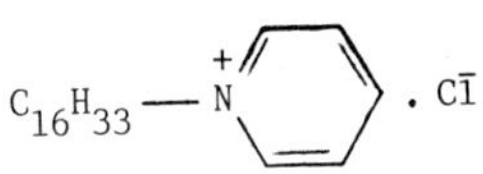	0.37	Hazy
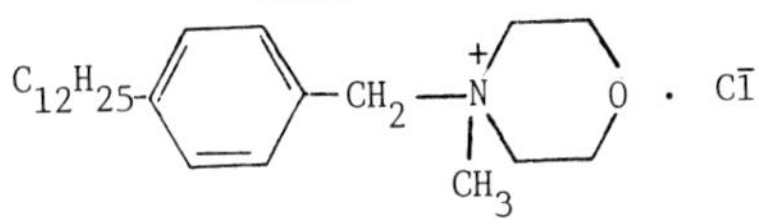	0.27	Clear

The importance of molecular weight and solvent solubility is also illustrated with the following series of amide compounds (Table 5).

Table 5. The resistance of xylene containing 4% amide additive

	Resistance MΩ	Solubility
$C_{11}H_{23}CONH-(CH_2)_3-\overset{+}{N}(CH_3)_2C_2H_5 \cdot C_2H_5S\bar{O}_4$	8	Cloudy
$C_{17}H_{35}CONH-(CH_2)_3-\overset{+}{N}(CH_3)_2C_2H_5 \cdot C_2H_5S\bar{O}_4$	0.13	Clear
$C_{17}H_{35}CONH-(CH_2)_3-\overset{+}{N}(CH_3)_3 \cdot (CH_3)_2P\bar{O}_4$	0.14	Clear

Choosing the best quaternary ammonium compound involved the investigation of various chemical structures, molecular weights and anions. Using commercially available blends of raw materials and avoiding potentially corrosive ions, it has been possible to develop a particularly effective cationic agent referred to as CA [1]. Bearing in mind the customer preference for pourable liquids, an 80% solution in butanol is normally used to give very low resistivities with small additions, as illustrated with xylene in Figure 3.

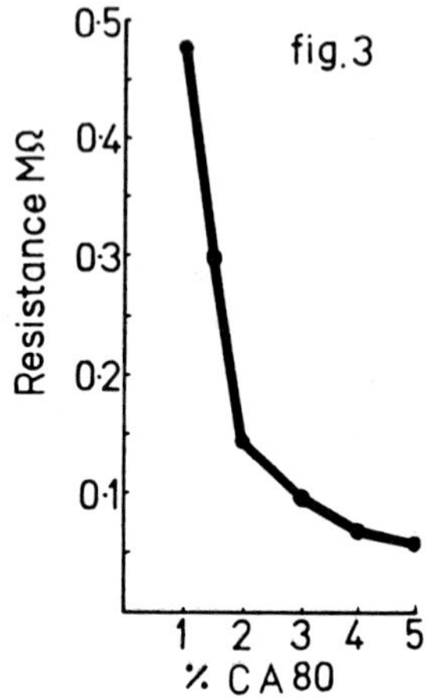

When using this product in paints it is advantageous to add a modest amount of polar solvent because a synergistic effect is usually observed (Table 6).

Table 6. Effect of CA in a long oil alkyd paint.

	Resistance M Ω
Paint + Hydrocarbon Solvent	20
Paint + Hydrocarbon Solvent + 1% CA80	1.8
Paint + Hydrocarbon Solvent + 10% Methyl Ethyl Ketone + 1% CA80	0.29

To summarise the advantages of using CA as a conductivity additive:-

1. Compatible with most types of paint thus avoiding reformulation.
2. The fluid 80% solution is quickly and easily dissolved.
3. Only low dosage is needed to produce a large drop in resistivity (0.5 - 3% in practice).
4. The same low dosage will reduce surface tension to a suitably low value.
5. Yellowing effects are not usually observed even in sensitive white paints.
6. Minimum effects on the paint film properties such as evenness, general appearance, hardness, corrosion resistance etc.

Antistatic Additives for Plastics, in particular Polyurethane.

The many adverse effects of static electricity on insulating plastic materials are well recorded [2,3]. These include textile processing difficulties, poor sound reproduction from records, excessive dust and dirt collection on surfaces and dust or solvent explosion hazards. One of the most important hazards associated with modern technology is electrostatic discharge (ESD) which can cause critical damage to microchip circuits, so essential in modern electronics. Some examples of the electrostatic voltages which a person can generate without any awareness are illustrated in Table 7 [4], at two levels of relative humidity.

Table 7. Typical electrostatic voltages

Generation	Voltages 10-20% RH	65-90% RH
Walking across carpet	35,000	1,500
Walking over vinyl floor	12,000	250
Working at bench	6,000	100
Pick up polybag	20,000	1,200
Chair padded with polyurethane foam	18,000	1,500

ESD's of 500 V can damage electronic components if discharge to earth occurs as a pulse of electrical energy. In fact discharge to earth is not essential because being placed in the electrostatic field near a charged object may induce sufficiently high damaging potentials across sensitive components.

Various types of antistatic agent are reviewed in the literature [2] and it appears that surfactant molecules are particularly effective. Non-ionic, anionic, cationic and even ampholytic types have been used as internal antistatic agents, depending on the type of polymer. For example, non-ionics are recommended for polyolefins but cationics are particularly effective in the more polar polymers [3] such as PVC, polystyrenes and polyurethanes. Polyurethanes are of particular interest in the three major forms:-

Elastomer	-	Solid polyurethane with no foam structure, e.g. conveyer belt pulleys in coal mining and processing.

High Density Foam - a polymer with a compact foam structure, e.g. safety shoe soles for use in industry, operating theatres, electronic clean rooms.

Low Density Foam - a highly expanded polymer used for packing electronic components and equipment.

The large influence of relative humidity on the effectiveness of antistatic agents suggests that a film of water is held on the plastic surface by the presence of the ionised or polar groups of the surfactant molecular projecting from the surface, with the hydrophobic tail anchored in the polymer.

Measuring surface resistivity (Figure 4) offers therefore a method of determining the effectiveness of the antistatic agent. Conditioning the sample at say 50% relative humidity for many hours is essential and even then it is difficult to obtain reproducible results.

fig. 4

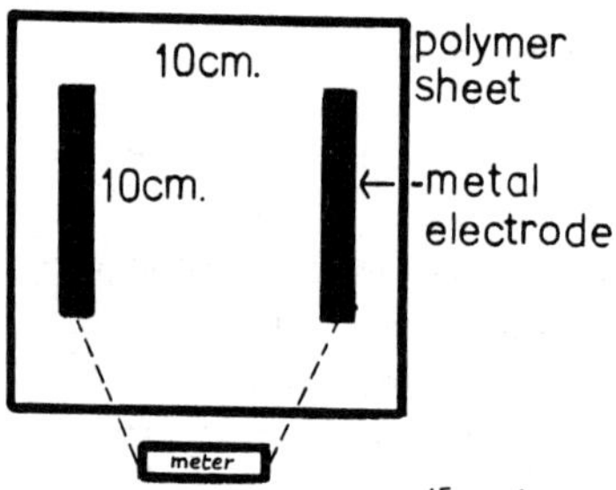

insulator $>10^{15}\,\Omega/\mathrm{sq}$

fig. 5

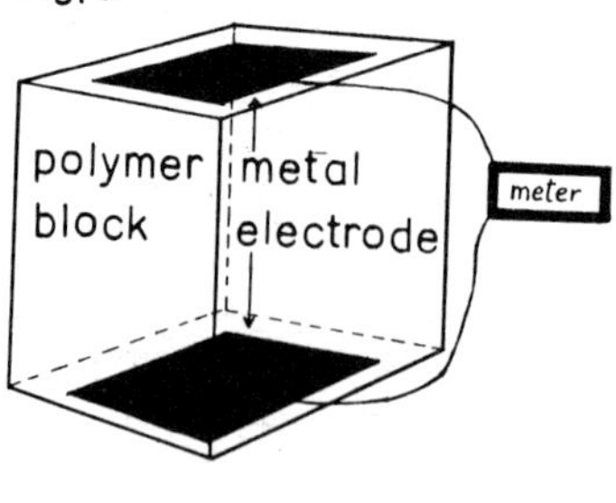

insulator $>10^{12}\,\Omega.\mathrm{m}$

Volume conductivity is therefore often measured to give a more reproducible value (Figure 5), although specifications may demand test methods which simulate use conditions. Resistance between an inner wet electrode and an outer dry electrode across a shoe sole is measured in BS2050 which requires a volume resistance below 5 x 10^7 ohms or in the German DIN4843 which requires a value below 10^8 ohms, under the particular conditions of the test.

Generally recognised levels of surface resistivity [4] for the various types of Polyurethane are shown in Figure 6.

Figure 6. Surface resistivity spectrum

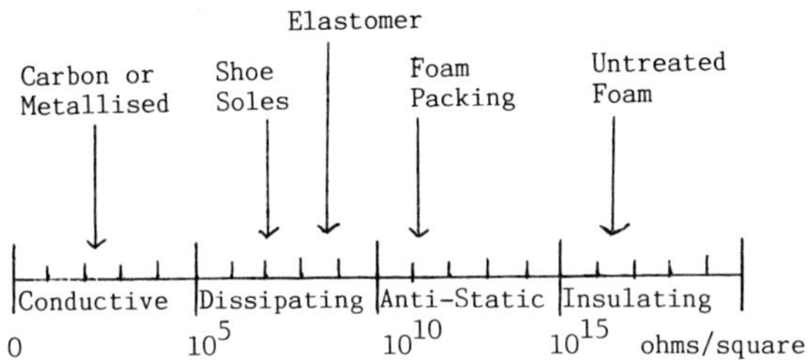

Very conductive plastics are usually made by filling with large quantities (10 - 30%) of carbon or powdered metal to produce a continuous conductive path through the polymer bulk and it is unlikely that organic additives, relying on moisture absorption, can ever produce such low resistivities.

A large variety of quaternary ammonium compounds have been tested in various different polyurethanes and the same blend represented by CA was the most effective material, conveniently used as an 80% solution in butane-1,4-diol referred to as PU [1].
In this form it is easily dissolved in the polyol component at a level of 2 - 4% on total weight of polymer. Furthermore, the dihydric alcohol does not act as a chain terminating agent for the following polymerisation reaction [5].

OCN - R - NCO + HO - Polyester - OH
diisocyanate (HO - Polyether - OH)

↓ Catalysts, Tertiary amine and Stannous octoate

—— O - C(=O) - HN - R - NH - C(=O) - O - Polyester - O - C(=O) ——

Polyurethane

Foam generation occurs by reaction with a small water addition to form carbon dioxide bubbles.

$$R{-}NCO + H_2O \longrightarrow [R{-}NH{-}\underset{\underset{O}{\|}}{C}{-}OH] \longrightarrow R{-}NH_2 + CO_2$$

The following examples are used to illustrate the variety of compounds tested.

Elastomer. National Coal Board Specification 158, relating to conveyer belt systems, states a value of 3×10^8 ohms/square for maximum surface resistivity. Table 8 illustrates the effectiveness of several different quaternary ammonium compounds in this type of material.

Table 8. Surface resistivity of elastomer with 2% additive

Antistat	Surface resistivity ($\times 10^8$ ohms/sq.)	
	Initial	After 1 month
PU	1.4	1.6
$C_{11}H_{35}CONH\text{-}(CH_2)_3\text{-}\overset{+}{N}(CH_3)_2C_2H_5 \cdot C_2H_5S\bar{O}_4$	1.3	2.6
$C_{17}H_{35}CONH\text{-}(CH_2)_3\text{-}\overset{+}{N}(CH_3)_2C_2H_5 \cdot C_2H_5S\bar{O}_4$	1.6	3.5
$C_{17}H_{35}CONH\text{-}(CH_2)_3\text{-}\overset{+}{N}(CH_3)_3 \cdot (CH_3)_2P\bar{O}_4$	3.7	2.8

The compound containing a phosphate anion produced unacceptable changes in mechanical properties of the plastic because it exerted an unwanted catalytic effect. Time dependent changes in resistance can occur, probably because the surfactant molecules can migrate both into and out of the critical surface layer [3].

High Density Foam. Resistance specifications for safety shoe soles, BS2050 and DIN4843 (German), have been achieved using PU. These special test methods require a maximum resistance of 5×10^7 ohms and 10^8 ohms respectively. Typical values using a variety of compounds are illustrated in Table 9.

Table 9. Resistance measurements on high density polyurethane foam containing 3% additive

Antistat	Resistance measurement ($x10^7$ ohms) Initial	After 7 days
Control (no additive)	> 20.0	> 20.0
$C_{12}H_{25}-C_6H_4-CH_2-\overset{+}{N}(C_2H_5)_3 \cdot C_2H_5S\bar{O}_4$	> 20.0	14.8
$C_{14}H_{29}-\overset{+}{N}(CH_3)_3 \cdot B\bar{r}$	> 20.0	10.6
PU blended with 20% $C_6H_5-CH_2-\overset{+}{N}(C_2H_5)_3 \cdot C\bar{l}$	18.0	7.2
$C_{10}H_{21}-\overset{+}{N}(CH_3)_2C_2H_5 \cdot C_2H_5S\bar{O}_4$	9.5	6.1
PU	14.0	3.3

Once again it was observed that a phosphate anion such as $(CH_3)_2P\bar{O}_4$ catalysed the foam formation, resulting in an unsatisfactory foam structure and consequently poorer mechanical properties.

Bearing in mind the possible importance of migration of the surfactant molecules in the antistatic mechanism, it was decided to investigate the effect of using the following reactive quaternary ammonium compounds.

$$C_{18}H_{35}-\overset{+}{N}(CH_2CH_2OH)_2-C_2H_5 \cdot C_2H_5S\bar{O}_4 \quad \text{and} \quad CH_3-\overset{+}{N}(CH_2CH_2OH)_2-C_2H_5 \cdot C_2H_5S\bar{O}_4$$

Polymerisation occurred normally and presumably the dihydric alcohols will react just like the normal polyol component to become chemically locked into the foam. The antistatic effect was poorer than that achieved with agent PU although it would be interesting to investigate the long term effects on a solid elastomer subject to a washing treatment, which might be expected to remove unfixed surface material.

Low Density Foam. The properties of packaging materials used for electronic components and sensitive electronic equipment are of particular interest to the Ministry of Defence and the appropriate American authority. At the present time, a volume resistivity of about 10^8-10^9 ohms. metre is desired together with a surface resistivity of 10^{11} ohm/square or less. A typical value of 1.8 x 10^{10} ohms/square was achieved using PU added at 5% on the polyol weight (2 - 3% on total weight). Increasing the molecular weight of the additive only produced a marginal improvement to 1.45 x 10^{10} ohms/square but the resulting semi-solid paste was much more difficult to handle and disperse into the polyol phase. Minimum corrosion characteristics are also required for this type of packing foam, so halide ions are not favoured.

Desirable Properties. The extensive testing of a large variety of quaternary ammonium compounds was designed to produce a product with the following properties.

1. Antistatic or ESD control at low concentrations, e.g. 2 - 4% on total weight.
2. Minimal effect on processing reactivity.
3. Minimal effect on other foam or elastomer properties.
4. Soluble and stable in the polyol component.
5. Easily handled, down to low temperatures, as a concentrated pourable solution.
6. Relatively high flash point.
7. Moderate toxicity.
8. Free from corrosive halide ions.

The cationic agent CA has in fact been sold in a variety of solvents to suit various customers e.g. butanediol, butyl oxitol, ethylene glycol and methylene chloride. Sometimes the normal solvent may not be entirely suitable for some particular foam structure. Our latest antistatic agent, referred to as PP50, is prepared as a 50% solution in TRICHLOROPROPYL PHOSPHATE, a compound which is often added at a level of 10-15% to polyurethane foams as a fire retardant.

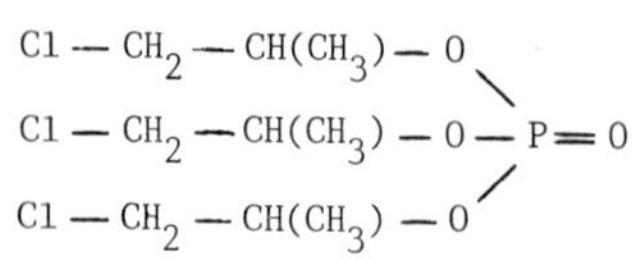

Advantages of PP50. This alternative material is proving very popular for the following reasons.

1. No hydroxyl groups: therefore it cannot participate in the polymerisation process to affect foam structure.
2. Very low moisture content and non-hygroscopic. It is particularly suitable for elastomer manufacture where the virtual absence of water is desirable.
3. Not inflammable.
4. Contributes a fire retardant effect.
5. If this type of fire-retardant is already added then some additive cost is saved.
6. A pourable liquid down to reasonably low temperatures.

Other Applications. The use of quaternary ammonium antistats can be restricted by the high temperatures reached in making or moulding plastic materials. However, when high temperatures exist only for short periods blend CA can be used in other plastics such as PVC, polystyrene, rubber etc. External application is also useful, when the effect is more temporary. It can be used on fibres and in the cleaning of surfaces, particularly in clean rooms of electronic component manufacturers, where its biocidal action is an added bonus.

Acknowledgements.

I wish to thank the following people for their assistance in making and evaluating many chemicals and for their help in the preparation of this paper:

paper:	A. Lerner	-	Aceto Chemical Co. (U.S.A.).
	B. Milner	-	A.B.M. Research Dept., Leeds.
	A.P. Pearson	-	A.B.M. Technical Services Dept., Leeds.

References.

1. ABM Chemicals Ltd., "Conductivity Additives".
2. Kirk-Othmer "Encyclopedia of Chemical Technology", Wiley, New York, (3rd Edition), Vol. 3, p. 149.
3. R. Gächter and H. Müller, "Plastics Additives Handbook", Hanser, Munich/Vienna/New York, (2nd Edition 1984), Chapter 12, p. 565.
4. A. Lerner, "A New Additive for Electrostatic Discharge Control in Foams and Elastomers", Proceedings of the Society of the Plastics Industry, Texas, 1984, p.331-334.
5. J.M. Buist and H. Gudgeon, "Advances in Polyurethane Technology", Maclaren, London.

Industrial Applications of Co-polymer Surfactants

By M. A. Plant
ICI PLC, PETROCHEMICALS & PLASTICS DIVISION, PO BOX 90, WILTON CENTRE, MIDDLESBOROUGH, CLEVELAND TS6 8JE, UK

Introduction

The idea of utilising the hydrophilic character of the polyoxyethylene chain and the hydrophobic character of the polyoxypropylene chain to produce a surfactant was first developed in the USA more than 20 years ago. Since then the copolymer surfactants made from ethylene oxide and propylene oxide have grown into a large family of exceptionally versatile products.

But first, it is useful to place this group of products into the context of the total surfactants spectrum.

Figure 1 shows the major surfactants which are presently utilised industrially, segmented into nonionic, anionic and cationic product ranges.

Figure 1. Major Industrial Surfactants

Nonionic	Alcohol Ethoxylates
	Alkylphenol Ethoxylates
	EO/PO Copolymer Alkoxylates
Anionic	Alkylbenzene Sulphonates
	Alcohol Sulphates
	Alpha-Olefin Sulphonate
	Alcohol Ethoxylate Sulphates
	Paraffin Sulphonate
	Ligno Sulphonate
Cationic	Quaternary Ammonium Salts
	Amine Derivatives
	Betaines
	Imidazolines

In the chemical structures of most surfactants the hydrophilic or hydrophobic entity is essentially fixed in character across that particular range of surfactants. Thus, for example, in the alkylbenzene sulphonate range, the hydrophilic sulphonate grouping remains fixed and the character of the surfactant is modified by changing the nature of the alkylbenzene unit.

Similarly with the alcohol ethoxylates, the same hydrophobic long chain alcohol unit is modified by altering the length of the hydrophilic polyoxyethylene chain. Of course, a step change to another alcohol substrate is possible.

With the ethylene oxide/propylene oxide (EO/PO) copolymer surfactants a continuum of hydrophilic and hydrophobic character is achievable, depending on the molecular ratio of the copolymerised monomers.

An analysis of the applications of surfactants in W Europe is given in Figure 2. Nonionic surfactants find application more widely in the industrial sector and the EO/PO copolymers follow this trend.

Figure 2. Applications of Surfactants (%). 1984 W Europe

	Detergents	Cosmetics	Industrial
Nonionic	48	6	46
Anionic	72	8	20
Cationic	72	1	28

W Europe's total surfactant demand analysed by "ionic category" is shown in Figure 3. Although the anionic surfactant group is largest in volume terms across all applications, the nonionic products have the largest tonnage usage in the industrial sector.

Figure 3. Surfactant Demand. 1984 W Europe

	Demand (Ktes)	% Share
Nonionic	398	32%
Anionic	740	59%
Cationic	106	9%

EO/PO Surfactants

As with all surfactants an understanding of chemical structure and nomenclature greatly facilitates the understanding of applications and capabilities; in the case of the EO/PO surfactants this is particularly true. Four basic groups of products will be described; the first is shown in Figure 4.

Figure 4. Poloxamer

$$H(OCH_2CH_2)_m\,(OCH(CH_3)CH_2)_n\,(OCH_2CH_2)_mOH$$

$$(EO)_m\,(PO)_n\,(EO)_m$$

The central molecular segment consists of a polymerised propylene oxide (PO) substrate onto which ethylene oxide (EO) is then block copolymerised giving the potential for a very large range of related surfactants (Figure 5).

There are clearly many ways to represent graphically the proportion of EO and PO in these surfactants; traditionally the molecular weight of the polyoxypropylene chain is placed on the vertical axis and weight % of copolymerised EO is represented on the horizontal axis. This actually represents the way the products are manufactured industrially.

To facilitate the use of the nomenclature, the axes are given arbitrary numerical co-ordinates and any product can be described in terms of these co-ordinates by two or three digits.

The last digit represents the weight % of EO and the first one or two digits the molecular weight of the polymerised PO hydrophobe. The preceding letter shows the physical character of the surfactant; liquid, paste or flake.

The enormous range of surfactant properties provided by these products is demonstrated in Figure 6.

Over thirty products are of industrial interest across this structure and property grid, varying in molecular weight from 1500 to 15000.

Figure 7 shows the HLB values (hydrophilic-lipophilic balance) of the Poloxamer family of surfactants. The products with HLB values below 6 tend to be hydrophobic whereas those with HLB values between 8 and 30 are essentially hydrophilic; again, the wide flexibility of potential application is clearly demonstrated.

Figure 5. Poloxamer Grid

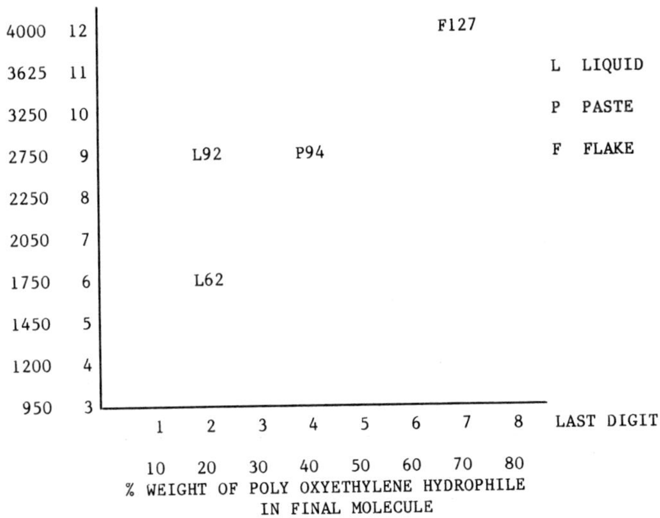

Figure 6. Poloxamer Grid. Properties

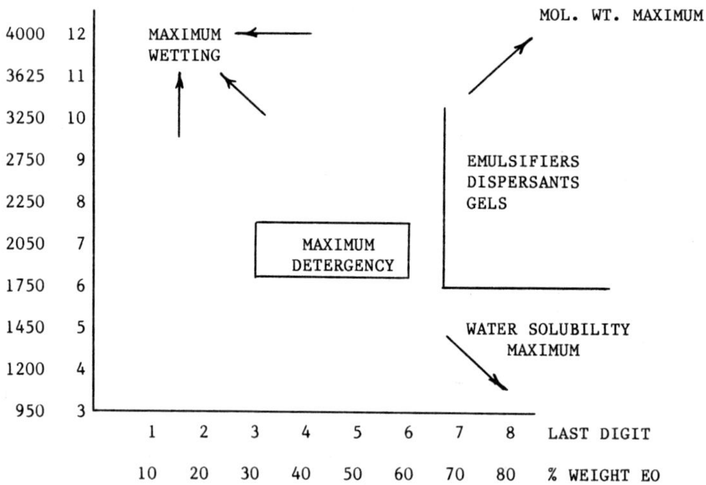

Figure 7. Poloxamer Grid. HLB Values

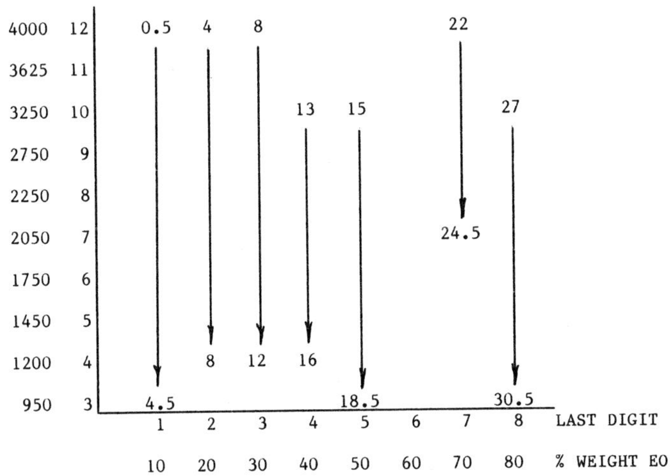

The discussion has so far centred upon molecules where the hydrophilic (EO) chain is copolymerised onto a hydrophobic (PO) substrate. The reverse structure is possible and exhibits interesting properties in certain applications; Figure 8 shows the structure and nomenclature of this range of surfactants. The external polyoxypropylene chains terminate with secondary hydroxyl groups which provide greater chemical stability, in alkaline media for example.

Figure 8. Meroxapol. Reverse Poloxamer

$(PO)_b(EO)_a(PO)_b$

25 2500 Total molecular weight of PO hydrophobe
R Reverse
2 20% weight of EO hydrophile in total molecule

Secondary OH terminal groups.

Ethylene diamine based products

The Poloxamer type products can be considered as dipropylene glycol onto which is polymerised propylene oxide followed by ethylene oxide. Any reactive starter molecule can be used and ethylene diamine based products have proved interesting industrially (Figure 9). It is believed that the presence of two nitrogen atoms imparts modest cationic character to the Poloxamine series of products.

Figure 9. Poloxamine.

$$\begin{matrix}(EO)_y(PO)_x \\ (EO)_y(PO)_x\end{matrix} > NCH_2CH_2N < \begin{matrix}(PO)_x(EO)_y \\ (PO)_x(EO)_y\end{matrix}$$

also Minoxapol or Reverse Poloxamine

Figure 10 shows examples of the Poloxamine series displayed on a grid; the trends of physical and surfactant properties of these products mirror the Poloxamer series.

About ten Poloxamines find major industrial application and the molecular weights of the products spread widely between approximately 1500 and 25000.

Figure 10. Poloxamine Grid

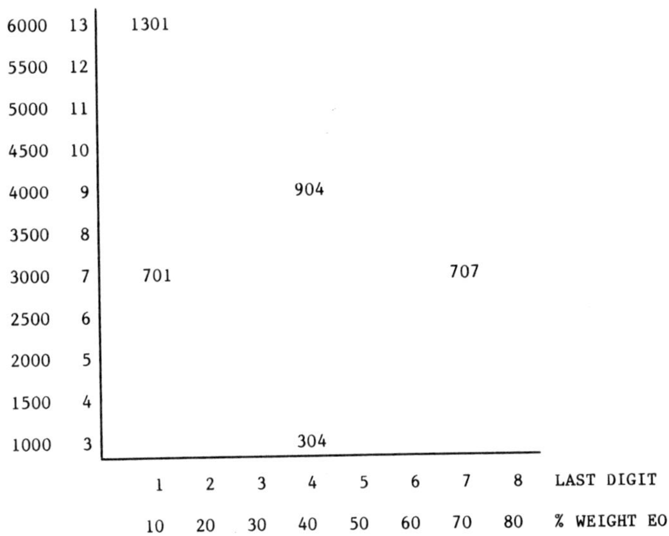

<u>Alcohol based products</u>

If the Poloxamer series can be considered to start from a diol, many other alcohol substrates are possible and indeed do offer products with attractive industrial properties (Figure 11). Copolymers based upon synthetic C_{13-15} alcohol have found wide application (Figure 12) and in this range of surfactants the concept of block and random copolymerisation is utilised.

Figure 11. Alcohol Derivatives

Alcohol - $(EO)_x/(PO)_y$	Block Reverse Block Random

Long chain monohydric alcohol.

Diol. eg. Poloxamer
eg. Meroxapol

Triols. Glycerol
Trimethylolpropane

Polyols and Sugars.

The block alcohol EO/PO copolymers are produced in the same way as the Poloxamer or Poloxamine series surfactants by first copolymerising propylene oxide onto the C_{13-15} alcohol and then further reacting ethylene oxide. Reverse products are again possible; EO first then PO.

Figure 12. C_{13-15} Alcohol - $(EO)_x/(PO)_y$

Block or Random

	High EO	Low PO	Low foam non-gelling detergents
Alcohol +			Low foam wetting agents
	Low EO	High PO	Antifoam/defoam with wetting properties

The random alcohol EO/PO copolymers are produced by first mixing the two oxides in a premixing vessel and injecting the blend onto the alcohol substrate for reaction. The degree of "randomness" is debatable as the ethylene oxide tends to react faster than the propylene oxide; in this case products very similar to the reverse alcohol EO/PO series may be expected.

Nevertheless the block, reverse block and random alcohol EO/PO copolymers do exhibit interesting property differences. Figure 12 demonstrates the general trend of properties in these three groups of products.

The alcohol EO/PO surfactants can be considered as simple alcohol ethoxylates in which the hydrophobic character is "tuned" by the block or random copolymerisation of propylene oxide.

Applications

The versatility of application provided by the EO/PO copolymers is shown in Figure 13. One of the strengths of the product series is the ease with which a property requirement can be interpolated into the spectrum by either blending or producing a new surfactant which fulfils the application need.

Figure 13. Summary of General Properties/Applications of Poloxamer, Poloxamine and Alcohol EO/PO products

Highly Versatile :-

Antifoamers	Emulsifiers
Defoamers	Emulsion Stabilisers
Wetters	Dispersants
Rewetters	Demulsifiers
Rinse Aids	Antistats
Degreasers	Viscosity Modifiers
Detergents	Antidust

Blends or tailor-made products for specific applications.

Some further comments on the Poloxamer and Poloxamine series specifically are given in Figure 14. The low toxicity of the Poloxamer series products has been clearly demonstrated by, for example, the use of F68 in intravenous pharmaceutical prepartions.

Interest in the use of these low toxicity products will no doubt be further stimulated by their inclusion in the listings of the US National Formulary (NF 16 XVI).

The additional stability of the terminal secondary hydroxyl in the reverse Poloxamer series is utilised in foam control applications for example. Aggressively alkaline cleaning or detergent environments, such as those required in some bottle washing or metal treatment applications, demand stability from the antifoam in the formulation and this is provided by the reverse Poloxamer structure.

Figure 14. Properties. Poloxamer, Meroxapol & Poloxamine series

Poloxamer	Low toxicity. Higher molecular weight - lower toxicity. Molecular weight range 1500-15,000
Meroxapol	PO Secondary Hydroxyl Terminal Group Improved alkaline stability
Poloxamine	N Atoms. Slightly cationic. Improved thermal/oxidative stability. Molecular weight range 1500-30,000

The slightly cationic character of the Poloxamine series is thought to contribute to their efficiency in dispersing clays and muds in oil field applications where products such as 904 and 908 are used.

Figure 15 compares the alcohol EO/PO series with the simple long chain alcohol ethoxylates. One of the major applications for alcohol EO/PO products is in the dishwashing area where foam control is required in the washing cycle and wetting agents, with low foaming characteristics, are required during the rinse cycle.

The alcohol EO/PO co-polymers also provides low foam detergents for industrial cleaning applications. Thus frequently in this series the copolymerisation of propylene oxide can be considered to "tune" the surfactant activity of the simple alcohol ethoxylate.

Figure 15. Properties of alcohol EO/PO copolymers

Alcohol - EO or Alcohol-(EO/PO) ?

PO allows "tuning" of hydrophile/hydrophobe properties

PO extends application range : detergents/wetters/antifoams

Lower pour points with PO

Lower tendency to gel with PO

Molecular weight normally less than 1500

Figures 16-19 list some typical industrial applications of the three series of products by considering the duty they are performing. The uses are specifically not given in any particular order of importance or volume of application.

Figure 16. Typical Industrial Applications of Poloxamer, Poloxamine and Alcohol EO/PO copolymers

<u>Emulsifiers/Dispersants</u>

Agricultural Chemicals
Paints and Pigments
Emulsion Polymerisation/Plastics, Rubber
Leather Production
Pharmaceuticals
Metal Working Fluids
Drilling Muds/Oil Field Chemicals
Cosmetics
Water Treatment

Products of the Poloxamer and Meroxapol type are used in the widest number of applications and comprise the largest tonnage worldwide of the three series of products.

Figure 17. Typical Industrial Applications of Poloxamer, Poloxamine and Alcohol EO/PO copolymers

<u>Antifoamers/Defoamers</u>

Sugar Production
Pharmaceuticals Production
Fermentation
Paper Manufacture
Water Treatment
Industrial Detergents
Metal Cleaning
Starch Adhesive Production
Gas Treatment
Drilling Muds/Oil Field Chemicals
Metal Working Fluids
Dishwashing
Dishrinsing

Alcohol EO/PO copolymers have found major application in the foam control and wetting areas for domestic and industrial dishwashing. Their use spreads very widely in other industries however.

Figure 18. Typical Industrial Applications of Poloxamer, Poloxamine and Alcohol EO/PO copolymers

<u>Detergents</u>

Hard Surface Cleaners
Metal Degreasing (alkaline)
Metal Pickling (acid)
Textile Scouring
Paper Industry
Dishwashing

<u>Demulsifiers</u>

Oil/Water Emulsion Breakers
Tank Cleaning and Desludging

Figure 19. Typical Industrial Applications of Poloxamer, Poloxamine and Alcohol EO/PO copolymers

Wetters

Agricultural Chemicals
Paper Industry
Leather Industry
Aluminium Anodising
Paints
Metal Cleaning
Foundry Applications
Secondary Oil Recovery
Dishrinsing

The uses of Poloxamines and Minoxapols vary almost as widely as the Poloxamer series. However, the cationic character and oxidative stability of the ethylene diamine derived products fit them particularly for tough industrial applications. Severe antifoam, emulsification and demulsification applications are currently in the range of major outlets for these materials.

Present and Future Developments

The flexibility of composition offered by EO/PO copolymers inevitably results in continuing change as technical and commercial requirements develop.

Some techniques of modification which have been studied and which continue to bear fruit are shown in Figure 20.

Partial or total esterification of terminal hydroxyl groups of many of these products with long chain acids results in the modification of the hydrophobic character of the basic copolymer. Products of this type find application as foam control additives for fermentation, water treatment and sugar processing for example.

Figure 20. Present and Future Developments

Infinite range of permutations/combinations:-

Modify existing products by further reaction
eg esterify terminal hydroxyl
eg end capping

Alternative long chain alcohols in LF series

Alternative polyhydroxylic alcohols

Alternative alkene oxides to EO and PO

The chemical stability of the copolymers can be improved by "end capping" terminal hydroxyl groups with a variety of agents, thereby widening the potential range of application without significantly changing the surfactant character of the particular copolymer.

The polydentate structures offered by the use of polyhydroxylic starter molecules can provide interesting properties. Products based on trimethylol propane, glycerol or pentaerithritol offer useful characteristics in, for example, foam control and emulsification applications. The polydentate shape of the copolymer molecule produces different effects at the phase interface compared with a straight chain copolymer.

Various alkene oxides have been studied as alternatives to ethylene and propylene oxides; although some potentially useful property modifications are possible, the cost of alternative oxides precludes wide industrial interest at the moment.

Figure 21 shows some of the applications which are continuing to develop as market demands or technical capabilities change.

The question of biodegradability has been of particular interest recently as, in 1989, EEC regulations will increase the number of application areas where biodegradable surfactants are required. Again the modification of the EO/PO copolymers to produce the required properties is proving possible.

Figure 21. Some Areas of Current Development

Antifoams - fermentation, sugar etc

Wetters and Detergents - aggressive media

Oil Field Chemicals

Biodegradability - bottle washing
- metal treatment
- home and industrial dishwashing
- industrial cleaning

Summary

The ethylene oxide/propylene oxide copolymer surfactants are an exceptionally versatile range of products which find application in an equally diverse group of industries. The original ranges of copolymer products, together with the more recent developments from them, will continue to provide answers to many questions in industrial surfactant applications.

References

ICI Petrochemicals & Plastics Division Literature; "Synperonic" PE, "Synperionic" T, "Synperonic" LF and "Ukanil" Data Sheets.

Flurochemical Surfactants

By M. C. Allison
3M (UK) LIMITED, 3M HOUSE, PO BOX 1, BRACKNELL, BERKS. RG12 1JU, UK

What is a Fluorochemical?

Chemically a fluorochemical is a hydrocarbon where all or part of the carbon-bonded hydrogen has been replaced with fluorine.

The amount of fluorine in nature is large. One of the largest types of rock structure is made from a mineral called fluorspar. There is little evidence that any form of organic fluorine containing material is found in a natural state, which is why organic fluorochemicals are an entirely man made class of compound.

Let us look at the impact fluorine has made on history. 1771 saw the discovery of hydrogen fluoride and it was 1886 before elemental fluorine was isolated. The next forty years or so saw steady progress until in 1928 the chloro-fluoro gases based on methane and ethane began to replace ammonia as refrigerant gas. The next leap forward required the intervention of the Second World War. Research in the USA was looking for compounds that could be of use in the armament race and the invention of Prof. Simon in 1937 of an electrochemical method for substituting fluorine for hydrogen saw the beginning of the use of fluorochemicals designed to be of use to industry. In 1945 Prof. Simon's patent was purchased and amongst many applications the introduction of fluorine based surface-active chemicals found its way to the market place; this took another six or seven years of research before reliable manufacture of products was possible.

Let us look at some definitions:

A fluorocarbon is a compound that contains no carbon bonded hydrogen, only carbon-fluorine bonds and possibly sulphur, nitrogen or oxygen.

A fluorochemical is a compound that has a fluorine containing group at one end and an organic carbon-hydrogen part at the other.

A fluorochemical intermediate is a fluorine containing compound that can be further reacted to produce a functional fluoro-chemical.

The substitution of fluorine for hydrogen in an organic molecule has some remarkable effects. If we take methane as a start, by substituting one fluorine atom for one hydrogen atom we find that the bond length increases to about 1.38 angstroms; this is because the Van der Waal's radius of fluorine is 1.35A against hydrogen at 1.2A. Further substitution of hydrogens results in a shortening of the carbon-fluorine bond length to about 1.32A. We find that fluorine fits quite well in place of hydrogen without causing any steric hindrance or strain on the carbon-carbon bonds. This is not the same for the other halogens. Even chlorine will produce sufficient strain to induce instability in the molecule as more halogen is added.

At the same time as shortening the bond length with increasing fluorination, there is an increase in the bond strength from 107 to 116 kcals per mole. This implies that the more heavily fluorinated a molecule is, the more stable it becomes. Along with this stability is the effect that increasing fluorination in a molecule such as acetic acid also increases the reactivity of the molecule. Acetic acid in its natural state is a relatively weak acid; by substituting fluorine in the methyl group there is an increase in dissociation until perfluoro-acetic acid is completely dissociated in aqueous solution at room temperature.

Dissociation Constants for Fluorine Substituted Acetic Acids.

Acid	Ka (H_2O at 25°C)
CH_3COOH	0.0000177
CH_2FCOOH	0.0022
CHF_2COOH	0.057
CF_3COOH	0.59

Figure 1

Fluorine being the most electronegative of the elements withdraws electrons to itself away from the acid hydrogen at the other end of the molecule.

What are the features of interest when we introduce fluorine into a molecule?

A completely fluorinated molecule such as perfluoro-octane (C_8F_{18}) is stable to organic reagents under normal conditions. Such a fluorocarbon will react with molten alkali or alkali-earth metals but with nearly everything else it is entirely inert. The boiling points of pure fluorocarbons are very similar to the noble gases of equivalent molecular weight. They have very low intermolecular forces which result in low boiling points compared to the analogous hydrocarbon, low heat of vaporisation and low surface tension. The electronegativity of the fluorine results in a material that has high dielectric strength and a low dielectric constant, an ideal medium for the direct cooling of electronic components and assemblies.

The solubility of perfluorocarbons is very low in both aqueous and organic liquids resulting in the immiscibility of fluorocarbons in both types of liquids.

Introducing perfluoro groups onto a molecule, or building a compound from a reactive fluorochemical intermediate, gives a whole range of commercially interesting products. More reactive herbicides and pesticides often result from the introduction of fluorine to the molecule and, as already stated, fluorine can be used to increase the activity of organic acids etc. Adhering the fluorochemical onto a solid surface will give water and oil/solvent repellency to many substrates such as textiles, leather, paper and even electronic printed circuits. By introducing a solubilising group onto a fluorochemical we have a material that is extremely surface active in solution. Fluorocarbons have very low surface tensions in the region of 15 to 18 dynes/cm and this property can be transferred to a solution of the fluorochemical, thus fluorochemical surfactants.

Surfactants based on fluorochemicals are analogous in structure to their hydrocarbon counterparts. The main difference is in the chemical and thermal stability of the surfactant and the extremely high efficiency, giving unparalleled performance at use levels as low as 25 ppm.

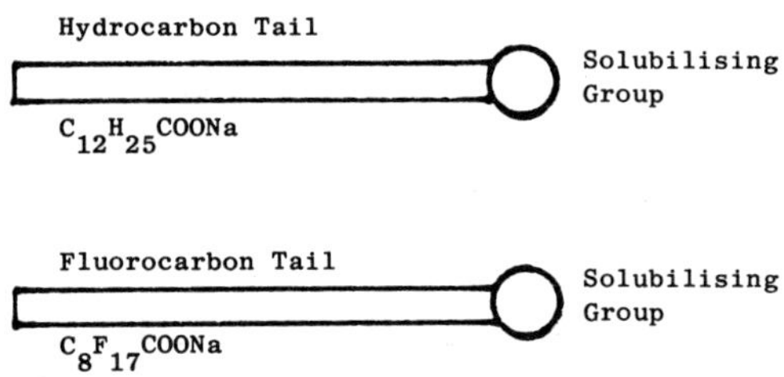

Figure 2.

The surfactants shown in Figure 2, simple sodium salts of carboxylic acids, will give in hydrocarbon form a minimum surface tension in water of about 28 dynes/cm at 10,000 ppm. The fluorocarbon analog gives about 18 dynes/cm at 50 ppm.

Fluorchemical surfactants come in the usual forms of anionic, cationic, nonionic and amphoteri. Their primary use is in reduction of surface tension of a given liquid.

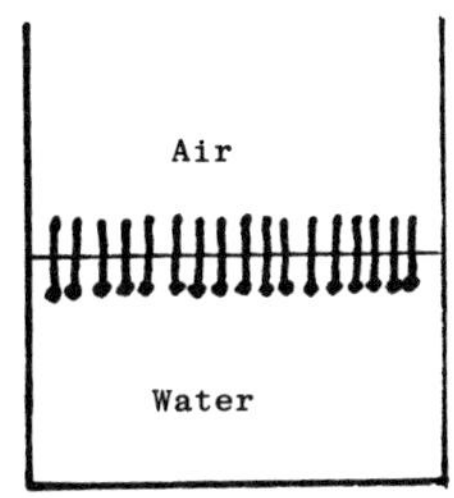

Figure 3.

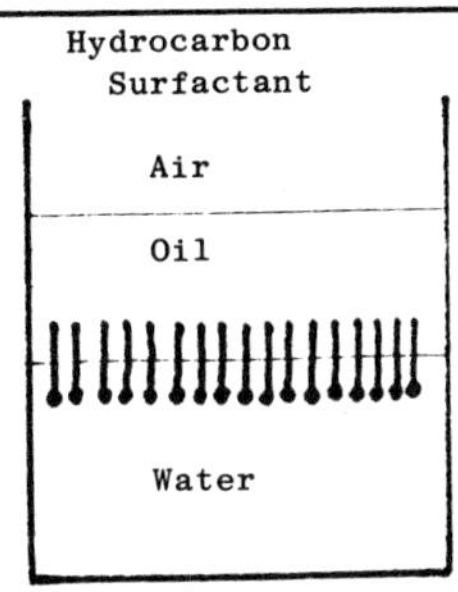

Figure 4.

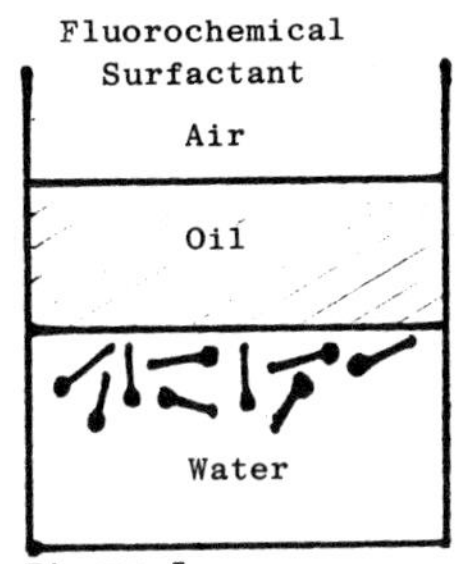

Figure 5.

Comparing fluorochemical surfactants with hydrocarbon surfactants shows that in aqueous solution (Fig. 3) the surfactant is oriented at the gas/liquid interface. When a two-phase system is studied (Fig. 4) such as hydrocarbon oil on top of water we find that the hydrocarbon surfactant is easily polarised at the interface of the liquids thus reducing the interfacial tension between the two. The fluorochemical surfactant is not attracted to the oil phase so there is little orientation at the liquid-liquid interface with consequent little change in interfacial tension. (Fig. 5) Thus fluorochemical surfactants are poor emulsifiers for water/oil systems, something which finds advantage in areas such as solvent extraction of minerals such as copper from acidic solutions. The surfactant is present to produce a stable foam blanket around the anode during the electro-winning process preventing bubbles from bursting and forming a toxic and corrosive acid mist in the tank house. If a hydrocarbon surfactant was used there is the danger of incomplete separation of the aqueous-solvent phases and subsequent loss of solvent chelatant.

One of the most important properties of fluorochemical surfactants is the large potential reduction in surface tension. This is important in the surface coatings industries when a liquid is required to wet and spread over a solid surface, often today one that has a low free surface energy such as PTFE or polypropylene. The parameters to examine are the surface energy of the solid, the surface tension of the liquid and the interfacial tension between the two. The spreading coefficient (S) is the surface energy of the solid minus the sum of the liquid surface tension and the interfacial tension.

Wetting and Spreading

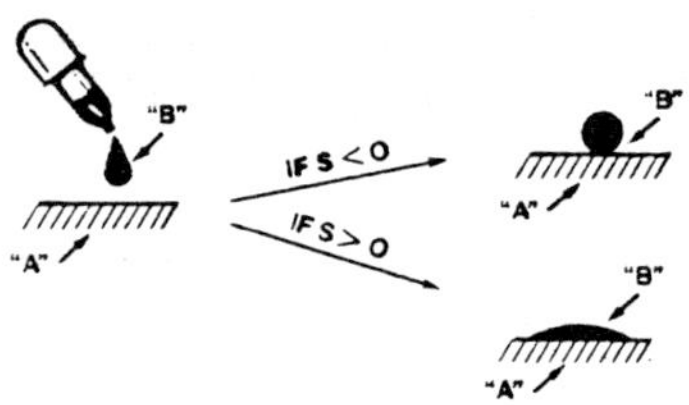

Spreading occurs only if S > O

Spreading Coefficient $S = \gamma_A - (\gamma_B + \gamma_{AB})$

Where
γ_A = Surface Energy of Substrate
γ_B = Surface Tension of Liquid
γ_{AB} = Interfacial Tension

Figure 6.

This coefficient must be positive for spreading to occur. For instance, if we take a wax surface with energy 25 dynes and attempt to wet it with a water based system using a surfactant, a fluorochemical surfactant would have to reduce the liquid surface tension to below 20 dynes before wetting and spreading would occur, because it is possible to have an interfacial tension as high as 5 or 6 dynes.

A hydrocarbon surfactant may well reduce the surface tension to 24 or 25 dynes but the interfacial tension will still be 1 or 2 dynes giving a negative spreading coefficient. It can be seen that a combination of fluorochemical surfactant at very low level could reduce the liquid surface tension 2 or 3 dynes below that of the solid and together with a hydrocarbon surfactant to reduce the interfacial tension to say 1 dyne we have wetting and spreading. Many examples of such synergy are exploited in industry.

Let us look at the differences between hydrocarbon and fluorochemical surfactants.

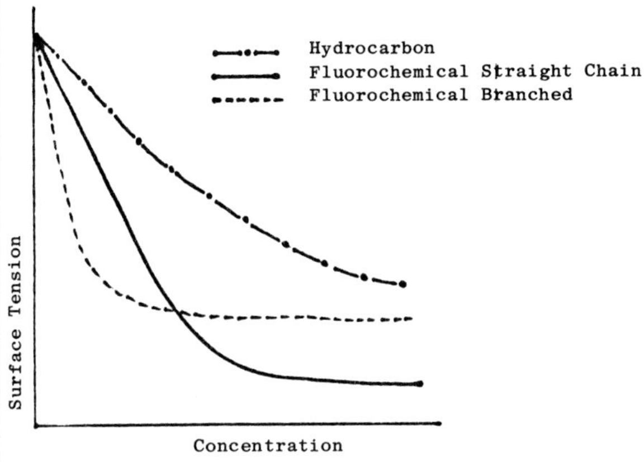

Figure 7.

Figure 7 shows that fluorochemical surfactants reach very low levels of ST at low concentration but also there is a difference between branched and straight-chain fluorocarbon groups. The branched chain surfactant fills up the liquid surface quicker than the straight chain: thus the ST reduces faster with increasing addition, but the straight chain surfactant can pack more molecules into the liquid surface and thus give lowest surface tension reductions.

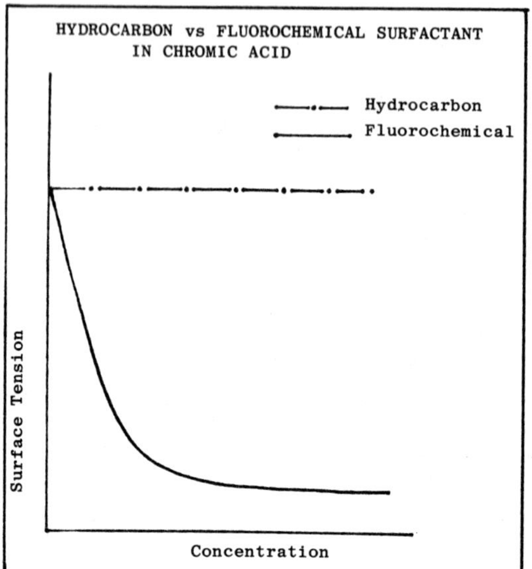

Figure 8.

The chemical and thermal stability of fluorochemical surfactants can be shown in this application in chromic acid (Figure 8). The hydrocarbon surfactant is completely destroyed by the chromic acid whilst the fluorochemical surfactant continues to be effective. This property is used in the decorative chrome plating industry to give a foam blanket on the liquid surface preventing bubbles from the electrodes bursting and forming a corrosive mist.

By using an organic soluble group on the surfactant we can product materials that are active in organic media. It is possible to reduce the surface tension of an epoxy resin from about 48 dynes to 20; this allows many coating and paint systems to have the wetting power required in almost any application.

Another surface coating application that can benefit from the use of organic active surfactants is in solvent based paints. Often coating problems such as orange peel, cratering and edge crawling arise from surface tension gradients produced in the coating during drying. These gradients are produced by differences in the surface tension between the solvent and the resin system, particularly when the solvent evaporates from thin parts of the coating. If the surface tension increases during solvent evaporation the bulk coating will flow towards the area of higher surface tension; if the reverse happens then the bulk liquid tends to draw in to itself. Selecting carefully the properties of solvent and resin can reduce this problem but a wider choice is available if a fluorochemical surfactant is added to the coating. Reducing the surface tension of the whole to a value below that of any of the components eliminates the introduction of ST gradients and their associated problems.

A summary of the benefits of organic active surfactants is their ability to wet low energy substrates; coat over contaminated substrates (hand prints from silicone creams etc); correct systems that have been contaminated (over-charge of defoamer etc) improve levelling and flow of the paint or coating.

Fluorochemical surfactants have been show to have utility in the following areas:-

Wetting for cleaners, coatings,etchants, rinse aids and inks.

Levelling of many coatings, finishes and polishes.

Foaming in chrome plating, anodising, solder flux, plastic foams and oil/gas well stimulation fluids.

Micellesin emulsion polymerisation of fluorocarbon monomers such as TFE and vinylidene fluoride.

Adsorption by some cationic surfactants have shown use in corrosion inhibitor formulations.

The concentration of surfactant normally used in these applications depends on the type of liquid system. For aqueous media the use range is about 50 to 200 ppm solids. Water reducible systems require 500 to 3000 ppm. Purely organic media would require as much as 1000 to 5000 ppm surfactant.

To summarise, fluorochemical surfactants offer the lowest possible surface tension, together with the lowest use concentrattion. They are extremely stable and active in many hostile environments and they are active in organic systems. The combinations of uses for these unique surfactants are still growing and will show benefits to industry for many years to come.

Subject Index

Absorption at mineral/water interface
   Collector adsorption 270, 276
   Electrical double layer 274
   Hemimicelles 276
   Potential determining ions 274
   Zero point of charge 275
Activators 282
Adhesion agents 266
Adhesive strength 29, 30
Admixtures
   Air-entraining 42
   Chemical 36, 37
   Concrete 33, 36
   Mineral 37
   Superplasticising 40
   Water-reducing 38
Adsorption 250, 254, 262, 266
Air entrainment 42
   Durability 45
   Effect on properties of concrete 44
   Frost resistance 46
Alcohol ether sulphates 66
Alcohol ethylene oxide - propylene oxide copolymers 326, 327, 329-331
Alkylbenzenesulphonates 63, 70
Alkylnaphthalenesulphonates 64
Alkyl amines, see amines, aliphatic
Alkyl amine acids 255
Alkyl betaines 255
Alkyl phenol ether sulphate 66
Alkyl sulphates (primary and secondary) 63
Alkyl sulphobetaines 255
Alkyl-o-xylenesulphonates 72
Amido amines 253
Amine collectors
   Commercially available 282
   Examples 271
   Ionisation 272
   Mechanism of adsorbtion 275
   Micellisation 273
   Solubility 272
Amines, aliphatic
   Alkoxylation 254
   Alkylation 254
   Carboxymethylation 255
   Chemical reactions 255
   Classification 261

Amines, aliphatic (cont.)
- Cyanoethylation 255
- Linear 251
- Properties 254
- Purification 253
- Surfactant behaviour 259
- Symmetrical 253
- Synthesis via nitrile 252
- Unsymmetrical, synthesis 253

Amines, ethoxylated 254, 262
Amphoteric compounds 255
- Mitigating effects on toxicity 263

Amphoteric silicones 25
Animal feed, nonionic surfactants for 191
Anionic silicones 25, 27
Anionic surfactants, history of development 61-65
Antifoam 48
- Solubility 51, 55

Antifoaming index 53
Antistatic additives for paint and plastics 307-317
Antistatic agents
- For textiles 203
- Other applications 317

Applications of latices 73
Application of organoclays 300
Aqueous film forming foam (AFFF) 91

Bacteriostatic properties of cationic surfactants 250
Basal spacing of clay organic complexes 293-296
Basicity of amines 254
Beidellite 292
Biodegradability 264
Biodegradability, nonionic surfactants 209
Bitumen emulsions 264-266
- Cut back 266

Branching of alkyl group in amines 251
- Effect on critical micelle concentrations 259
- Effect on melting point 256
- Effect on solubility 257

Breakthrough time 158, 163
Bridged amines
- Properties 254
- Synthesis 253
- Effect of bridging groups on melting point 256

Bubble size distribution 148, 155
Bulk polymerisation 78

Capillary force 128
Capillary number 128-131
Cationic surfactants
- Nitrogen-containing heterocyclic 251
- In mineral processing 269-288
- Worldwide consumption 250

Caustic solution flooding 127
Cellulose deresination, nonionic surfactants for 184
Cement
- Alkyl aryl sulphonates, sodium salts 42
- Amino acids 42

Cement (cont.)
- Calcium lignosulphonate 39
- Carboxylic acids 42
- Citric acid, sodium salts 39
- Gluconic acid, sodium salts 39
- Glucose 39
- Hydroxycarboxylic acid, salts 39
- Lignosulphonic acid, salts 39
- Polysaccharides 39
- Portland, composition 34
- Sodium citrate 39
- Sodium tartrate 39
- Sodium p-dodecylbenzenesulphonate 42
- Sodium gluconate 39
- Sodium lignosulphonate 39
- Sodium salts of wood resins 42
- Sucrose 39
- Sulphonated melamine formaldehyde condensate, salts 40
- Tartaric acid, sodium salts 35
- Water/cement (W/C) ratio 35

Cement, hydration of 34
Channelling effect 148
Chemical structure of nonionic surfactants 183
Clays 290
Clay complexes
- With alkylammonium cations 293, 294
- With non-charged polar molecules 293

Claying, prevention of 121
Cleaning performance, factors affecting 209
Coal-water slurries (CWS) 235-249
- Advantages of 238, 239
- Applications for 239, 240
- Chemicals used in 241
- Dispersing agents for 242-244
- Calorific values 246
- Economics 246-249
- Market for 246-249

Collectors, classification of 270
Conductivity additives for paint and plastics 307-317
Copper electro-winning 337
Corrosion inhibitors 260
Cosmetics, nonionic surfactants for 190
CO_2 flooding 127, 148
Coalescence of primary particles 77
Concrete
- Admixtures 36
- Flowing 41
- Frost resistance 46
- Hardened 35
- Production statistics 33
- Rheology (workability) 34, 35, 40, 46
- Setting time 35, 46
- Strength 40, 41, 46

Cost of industrial latices 88
Critical micelle concentration (CMC) 137, 151, 264
- Aqueous solution 268
- Effect in emulsion polymerisation 80, 81

Critical micelle concentration (CMC) (cont.)
  Emulsion polymerisation below 77
  Polar organic solvents 258
Crude oil 48
Crude oil carry-over 48

Dehydroabietylamine 251
Depressants 282
Diamines 266
Dioctahedral structure 291
Dispersants for coal water slurries 242-245
Dyeing assistants 201, 202

EEC Directive, biodegradability 209
Elastomers and additives 314
Electrical double layer 137
Electrochemical fluorination 333
Electrostatic interaction 108
Electrostatic spray painting 307
Emulsifiable concentrates for plant protection 187, 188
Emulsion polymerisation
  Advantages of 79
  Compared to other types 78
  Disadvantages of 79
  Process 73
Ether amines 253
Ethoxycarboxylates 66
Ethylenediamine 324
EO/PO copolymers
  Applications 327-331
  Developments 331, 332
Expansion ratio 90

Fabric conditioners, rinse-applied 264, 265
Fats and oils, naturally occurring 251
  Approximate composition 252
Fire fighting foam
  Classification 90, 91
  Methods of application 91
  Operational aspects of application 92
Flammability 264
Flocculation 283
Flotation
  Correlation with adsorption and contact angle data 278
  Description 270
  Effect of chain length 277
  Effect of concentration 277
  Effect of pH 279
  Effect of structure 281
  Examples 283
  Selectivity 282
  With cationic surfactants 277
Fluid displacement efficiency 158-169
  Effect of chain length of alcohols 162-167
  Effect of pressure 158-162
Fluorocarbons 334
  Solubility of 335

Fluorochemical foaming agents 91
Fluorochemical intermediates 334
Fluorochemical surfactants 333-342
  Adsorbtion of 341
  Branched and unbranched types 339
  C-F bond length 334
  C-F bond strength 334
  Chromic acid (effect in) 340
  Concentrations in use 342
  Foaming of 341
  Interfacial tension 338
  Levelling 341
  Micelles 341
  Surface tension in organic liquids 340
  Surface tension in water 336
  Wetting 341
  Wetting and spreading 338
Fluorosilicone oil 53
Foam 48
  Aqueous 49
  Collapse 49
  Collapse profile 49
  Non-aqueous 49
Foam flooding 127, 147-172
  Heavy oil recovery 169-172
Foaminess 151, 152
  Effect of chain length of alcohols 151
  Effect of temperature 152
Foaming agents
  Evaluation in fire fighting foams 96
  Formulation for fire fighting foams 93
  Textile processing 203
Foaming of latex 83
Foam quality, effect of chain length of alcohols 153
Foam stabilisation 26, 27
Foam stability 153
Food, nonionic surfactants for 191
Free radical addition polymerisation 74, 75

Gas-oil separation 48
Gas sparging tube 52
Gel formation 298
Gibbs elasticity 148
Gravity drainage 155
Gravity force 169
Gravity override 148
Greases 300
Gypsum slurry 68

Hard surface cleaning 227
  Cleaning methods 228
  Concentration effect 231
  Detergency testing 230
  Foam cleaners 229
  Institutional cleaners, types 230
  Nonionic surfactants, detergency 231, 232
  Solvents 231

Heat sensitisation of latex 82, 83
Heavy oil recovery 147
High density foam and additives 314, 315
HLB value of silicone nonionics 28, 29
Hydrolysis of silicates 261
Hydrophile 2
Hydrophile-lipophile balance, HLB 260, 264, 266
Hydrophobe 2
Hydrophobic groups, in cationics 251, 265

Imidazoline compounds 251, 253, 254
Industrial surfactants, consumption 180
Inks 301
Interbubble gas diffusion 148, 155
Interfacial charge density 134
Interfacial tension 127-134
Interfacial viscosity 131-134
Insecticidal properties 250
Interaction energy between particles 111-115
Iodonium compounds 250
Iron ore flotation 283

Kaolinite 293
Kier bleaching 201
Krafft effect 257

Latex product properties, surfactant effects 86
Laundry detergents 220
  Detergency 222, 223
  Kubelka-Munk equation 226
  Liquid formulations 225
  Storage stability 227
  Surfactant types, detergency 222, 223
  Surfactant types, foaming 221
  Testing, detergency 225
Levelling agents 204
Lignosulphonates 61
Liposomes in pharmaceuticals 189
Low density foam and additives 316
  Desirable properties of additive 316
Low surfactant latices 88
Lubricants 202
  For woollen spinning 202, 203

Machine dishwashing 211
  Defoaming 214
  Detergency measurement 212
  Detergency test, ASTM method 213
  Foaming, test method 214
  Food soil 214
  Rinse aid 215
  Rinse aid surfactants, dispersion 216, 218
  Rinse aid surfactants, foaming 218
  Rinse aid surfactants, surface tension measurement 217
  Rinsed articles, appearance 215, 216
  Rinsed articles, drying times 216
Manufacture of organophilic clays 298

Mass transfer of surfactant 134
Mechanism of gel formation 299
Mechanisms of particle formation 75-78
Meroxapol 324
Micellar mechanisms of particle formation 75-77
Microemulsions 136, 137, 143, 144
Microemulsions for plant protection 188
Milk replacers 192
Mobility control 134, 148
Monomer droplets 74, 75
Montmorillonite 292, 293
Morpholine compounds 251

Naphthenates 64
Nonionic silicones 26-28, 30
Nonionic surfactants
  Composition 224
  Industrial applications of 179-193
  Industrial consumption 180
  Low foaming 211
  Mode of action 244, 245
  Polyelectrolytes versus ethoxylated nonionics 243, 244
Nontronite 292

Oilwell drilling muds 302
Olefin sulphonates (alpha, internal, and vinylidene) 65
Oligomer precipitation and particle formation 77
Optimal salinity 136, 137
  Effect of oil chain length 137
Optical transmission foam meter 50
Organic complexes 293
Organic media 24, 26, 27
  Lowering surface tension in 29-31
Organophilic surfactants 24, 26, 27, 29-31
Ostwald ripening
  Role of polymers in 122, 123
  Role of surfactants in 122, 123

Paint 300
Paint additives 26
Paper coating latex, wet strength 87
Paper industry, nonionic surfactants for 184
Paraffin sulphonates 65
Particle formation theories 74-78, 79, 85
Particle size of latex 84-86
Partition coefficient 137, 260
Pegmatite flotation 284
Perfluoroacetic acid 334
Perfluoro-octane 335
Petroleum derived amines 251
Petroleum sulphonates 64
Pharmaceuticals, nonionic surfactants in 189
Phosphate esters 17-21, 65
  Dispersants 20
  Hydrotropes 21, 22
  Structure 17-19
Phosphonium compounds 250
Phosphorus pentoxide 19

Plant protection, nonionic surfactants in 186
Plasterboard 68
Plastics and static electricity 311
Polar additives 298, 299
Poloxamer 320
Poloxamer grid 320-323
Poloxamine 324, 325
Polydimethylsiloxane 51, 56
Polymers, role in colloid stability 106, 107
Polyurethane testing techniques 312
  Types 312
Portable foam apparatus 51
Primary oil recovery 127
Primary particles 75
Production statistics for surfactants 179
Properties of latex products, surfactants effects 86
Protein foams 91
Protonation, degree of 261
Pulp and paper, nonionic surfactants for 184
Pyridine compounds 251

Quaternary ammonium compounds 254, 265
  As additives 308, 309
  Commercial 297

Resistivity spectrum 313
Rheology of suspensions
  Influence of polymers 115-118
  Influence of surfactants 115-118
Role of surfactant in emulsion polymerisation 74, 78

Salicylate 65
Scouring 199
  Cotton 200
  Removal of contaminants 199
  Wool 200
Scondary oil recovery 127
Settling of suspensions 118-122
Shrink resist agents 207
Silicone amphoterics 25
Silicone anionics 25, 27
Silicone cationics 25
Silicone glycol 51, 53
Silicone nonionics 26, 27, 28
Silicone surfactants 24-32
  Applications 26, 28-31
  Structure 24, 25
  Surface activity 26-31
  Synthesis 25, 26
Smectites 290
Softeners for textiles 206
Soluble salt flotation 285
Solubilisation of iodine 258
Solubility of amines and derivatives 254, 264
  Amine salts in aqueous ethanol 258
  Aqueous solutions 257
  Non-polar organic solutions 258
  Organic solutions, oligomerisation 258

Soluble oil 143-147
Solution polymerisation 78
Stability of latex
  Surfactant effects 81, 82
  Various applications 82
Steam injection 127, 148
Steric interaction 109-111
Substantivity 261, 265
Substitution in layer structures 292
Sulphated castor oil 61
Sulphonated fatty acids and esters 65
Sulphonium compounds 250
Sulphosuccinates 65
Sulphoxonium compounds 250
Surface activity in organic media 24, 26-28
Surface charge 261
Surface modification of fabrics 265
Surface tension
  Dialkyl quaternary ammonium compounds 260
  Effect of chain length of alcohols 150, 151
  Effect of temperature 151
  Fluorochemical surfactants in organic liquids 340
  Fluorochemical surfactants in water 336
  Foaming agents 150, 151
  Sodium dodecylbenzenesulphonates 259
Surface tension 26-28, 29-31
  Adhesives 29, 30
  Lubricants 30, 31
  Release agents 31
Surface viscosity 153
  Effect of chain length of alcohols 153
Surfactant applications in
  Agriculture 11
  Civil engineering 12
  Food industry 13
  Industrial & institutional cleaning 9
  Leather and fur 15
  Metal and engineering 12
  Mineral extraction and processing 14
  Paints, lacquers, and inks 15
  Petroleum and oil 14
  Plastics, polymers, rubber and resins 16
  Pulp and paper 13
  Textiles and fibres 10
Surfactant choice in emulsion polymerisation 80
Surfactant effects in industrial applications 59
Surfactant effect on wet properties of latex products 87
Surfactant-polymer flooding 127-150
  Classification 136
  Effect of alcohol 143
  Effect of salt 136-139
  For highly viscous oil 147
Surfactant-polymer-foam flooding 170
Surfactant-polymer incompatibility 136
Surfactant solubility 143
Surfactants, role in colloid stability 106

Surfactant statistics
European/USA consumption 1985 6
European/USA consumption, domestic vs. industrial, 1985 7
UK industry sector, percentage change, volume 1985/6 7
USA/Japan/Europe 1982 4, 5, 6
Western European consumption by product type 1985 7
Surfactant types
Amphoterics 3, 25
Anionic 3, 62-66
Cationic 3, 251
Nonionic 3, 179-194
Suspension concentrates 104-106
Suspension polymerisation 79
Suspensions
Application of polymers for control of settling 118-122
Application of surfactants for control of settling 118-122
Sweep efficiency 150
Swelling of organoclays 294-296

Tertiary oil recovery 127
Tetraphosphoric acid (TPA) 18
Textiles
Auxiliaries for cotton and wool blends 197
Auxiliaries for synthetics 198
Auxiliaries for wool and wool blends 196
Processing aids 195
Theories of particle formation 74-78
Thermal combustion 127
Toxicity of cationics 261, 264
Acute 262
Trichloropropyl phosphate solvent 316
Trioctahedral structure 291

Unsaturation of alkyl group in cationics 251
Effect on melting point 256
Effect on solubility 257

Van der Waals Interaction 107-109
Vermiculite 293
Viscose industry 186
Viscosity of latex 85, 87
Viscous force 128

Walboard manufacture 68
Waste paper deinking 185
Water flooding 127
Water proofing for textiles 207
Water reduction 40
Water sensitivity of latex products 87, 88
Wettability at rock/fluid interface 134
Wetting angle 27
Wetting of fluorochemical surfactants 338, 341
Wetting speed 27
Wetting of substrates by latex 83

Zeropoint charge pH 261